A Gift for

_____

Presented by

_____

I Wish
I Knew
That
Math

I Wish I Knew That
Math

Cool Stuff You Need to Know

Dr. Mike Goldsmith

Reader's Digest

The Reader's Digest Association, Inc.
New York, NY / Montreal

A READER'S DIGEST BOOK

Copyright © 2012 Buster Books

All rights reserved. Unauthorized reproduction, in any manner, is prohibited.
Reader's Digest is a registered trademark of The Reader's Digest Association, Inc.

First published in Great Britain by Buster Books, an imprint of Michael O'Mara Books Limited,
9 Lion Yard, Tremadoc Road, London SW4 7NQ

FOR READER'S DIGEST
U.S. Editor: Barbara Booth
Consulting Editor: Jeanette Gingold
Designer: Elizabeth Tunnicliffe
Illustrator: Andrew Pinder
Senior Art Director: George McKeon
Editor-in-Chief, Books & Home Entertainment: Neil Wertheimer
Associate Publisher, Trade Publishing: Rosanne McManus
President and Publisher, Trade Publishing: Harold Clarke
Editor-in-Chief, North America: Liz Vaccariello
President, North America: Dan Lagani
President and CEO, Reader's Digest Association, Inc.: Robert H. Guth

Library of Congress Cataloging in Publication Data

Goldsmith, Mike, Dr.
   I wish i knew that. Math / Michael Goldsmith.
      p. cm. -- (Cool stuff you need to know)
   Summary: "With this book, kids can unlock the mysteries of math and discover the wonder
of numbers. From fractions to pi, and measurements to probability, kids will find out incredible
information, such as why zero is so useful; how to tell time on Earth and other planets; how music,
math, and space, are all connected and so much more!"-- Provided by publisher.
   Includes index.
   ISBN 978-1-60652-473-2 (hardback) -- ISBN 978-1-60652-474-9 (adobe) --
   ISBN 978-1-60652-475-6 (epub)
   1. Mathematics--Juvenile literature. I. Title. II. Title: Math.
   QA40.5.G65 2012
   510--dc23

                                      2012000709

We are committed to both the quality of our products and the service we provide to our customers.
We value your comments, so please feel free to contact us.

      The Reader's Digest Association, Inc.
      Adult Trade Publishing
      44 South Broadway
      White Plains, NY 10601

For more Reader's Digest products and information, visit our website:
      www.rd.com (in the United States)
      www.readersdigest.ca (in Canada)

Printed in the United States of America

1 3 5 7 9 10 8 6 4 2

CONTENTS

INTRODUCTION
GET INTO NUMBERS

Do fractions frustrate you? Do numbers drive you nuts? Never fear! This book will take you on a whirlwind tour through the fascinating world of math, until you're passionate about percentages and think decimals are divine!

It's not just about the mechanics of math, though. There are sections in this book about how math affects all kinds of things, from the way animals behave to the way you hear music. Lots of the greatest thinkers throughout history loved math and used it to invent cool stuff and discover new things.

Whether it helps you to tackle your homework or teaches you some new facts to freak out your friends, this book will make you a math lover, too—guaranteed.

NIFTY NUMBERS

ANCIENT NUMBERS

Long before calculators and computers were invented, if humans wanted to keep a record of things they had counted, they cut lines into sticks or bones. One of the earliest known examples of this kind of counting was discovered in a cave in South Africa. It was a baboon bone with 29 lines scratched into it. Tests show that the scratches were made about 35,000 years ago.

WATCH OUT— HE'S RUNNING OUT OF BONES!

Tallyho!

These lines, or tallies, may have been used to count anything from animals and people to passing days.

At first, the only number symbol used was 1. Actually, these were just scratches on bones, though, so if humans wanted to count to 1,000, they would have had to find a load of baboon bones and scratch 1,000 number 1s.

Today there are 10 different digits, or numerals—0, 1, 2, 3, 4, 5, 6, 7, 8, and 9. These digits make up what is called the decimal system—from the Latin word for 10: *decimus*.

The decimal system is a very logical way for humans to count—most people first learn about numbers when they start to count on their ten fingers. In fact, the word digit also means "finger." It's probably how you learned to count, and it's probably just how ancient humans started, too.

Count Like an Egyptian

The earliest known counting system based on the number 10 was used 5,000 years ago in Egypt. The Egyptians used sets of lines for numbers up to 9. They looked something like this:

Their new symbol for 10 was ∩, and larger numbers used combinations of ∣s and ∩s. So 22 was written: ∩∩∣∣ For 100 they used ℓ and for 1,000 ʃ, up to a million: ☥

A million seemed so massive to the ancient Egyptians that it also meant "any enormous number."

Timeless Numerals

The Romans also counted in tens using letters for numbers: I (1), V (5), X (10), L (50), and C (100). Later, D (500) and M (1,000) were added.

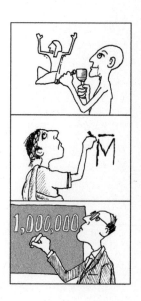

To write a number, letters were grouped together and added or subtracted according to the order. For example, if I is placed in front of a letter representing a larger number, it means "one less than." IX is 9, or one less than 10. The symbols CL were used to write 150—or 100 plus 50. So added together, the letters CCLVII stand for 257.

You see Roman numerals on some clocks or at the end of some TV programs, to show when they were made.

ALL ABOUT NOTHING

People had been using counting systems for centuries before they realized that something was missing—zero! Although the mathematician Ptolemy, in ancient Greece, did experiment with it, zero wasn't used regularly until the late ninth century.

Count Me In

Without zero, there is no way of telling the difference between, say, 166, 1,066, and 166,000. It is also a handy starting point for everything from stopwatches to rulers and temperature scales.

To tell the difference, a new counting system of "positional notation" was developed using the "place value" of numbers. This system divides numbers into columns, starting with ones, or units, on the right, with tens to the left, then 100s, then 1,000s, and so on. For example, with the number 3,975, you can easily see that there are three 1,000s, nine 100s, seven 10s, and five 1s.

In this system, after you reach 9, you place a 1 in the tens column and go back to 0 in the units column. At 19, the 1 in the tens column changes to a 2, and the units go back to 0 again, and so on, until you reach 99. Then a 1 is placed in the 100s column and the units and tens go back to 0.

HOW TO TALK TO COMPUTERS

The 10-digit number system is called base 10. However, there are other bases, too. The simplest is base 2, or "binary." It uses just two digits—1 and 0.

In binary, instead of writing "0, 1, 2, 3, 4, 5, 6, 7," and so on as normal, you would write, "0, 1, 10, 11, 100, 101, 110, 111" to represent the same numbers. This is because, just as in base 10, binary numbers can be thought of in columns. Instead of going up in 1s, 10s, 100s, 1,000s working from the right—increasing by a multiple of 10 each time—the value of the columns doubles each time. From the right, the first column is 1s, the next column to the left is 2s, the next is 4s, then 8s, then 16s, and so on. For example, the number 17 is written as 10001, which means: "one 16, zero 8s, zero 4s, zero 2s, and one 1":

16s	8s	4s	2s	1s
1	0	0	0	1

242?

NOW SAY IT IN BINARY.

This might not seem like a very useful way to count, but it is perfect for computing. Every computer is full of tiny electronic switches, each of which can either be on or off.

To a computer, a switch that is on represents a 1 and a switch that is off represents 0.

15

A set of switches in a computer can store a binary number. The number 5, for instance, would be stored like this:

Well, it would if the computer had elves inside it.

Computers use binary to store and work with all kinds of data, not just numbers. Everything from letters and sounds to pictures can be converted into binary code.

Did You Know?

There are many other bases as well as base 10 and base 2. Base 8, or "octal," and base 64 are also used in computing, as is base 16, or "hexadecimal," which is used to refer to areas within computer memories. It uses the digits 0, 1, 2, 3, 4, 5, 6, 7, 8, and 9 and the letters A, B, C, D, E, and F

TRICKY OPERATIONS

When you are counting or adding or subtracting, you are performing an "operation." Not the kind that doctors perform—the kind that mathematicians perform when they do arithmetic.

The word arithmetic comes from ancient Greek and means "the art of numbers." It is used in addition, subtraction, multiplication, and division, which are known as the four operations.

Line 'Em Up!

A number line is a good way to think about numbers and operations. Here is a number line showing the addition 2 + 2. The answer, known as the sum in addition, is of course 4:

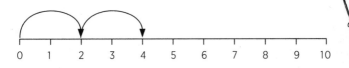

Subtraction is just as simple. To calculate 10 − 4, count backward along the number line from the first number, 10, using the second number, 4.

The answer is the difference between the two numbers. Here, it is 6:

Multiplication is repeated addition. For example, to calculate 3 x 4 using a number line, simply count along from 0 by the first "factor," or number, as many times as the second factor to get the answer. In other words, count to 3 four times! In multiplication the answer is known as the product. Here the product is 12.

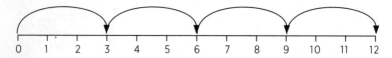

Division is repeated subtraction. The example below shows 6 ÷ 3. The number line is marked off from 0 to the dividend—the first number—then split into equal segments that match the second number—the divisor. The length of each segment is the quotient, which is what the answer is known as in division. Here, it is 2:

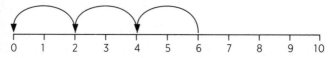

For addition and multiplication, it does not matter which way around the calculation is written: 2 + 3 is the same as 3 + 2. However, it's worth remembering that the same is not true for division and subtraction: 7 − 2 is not the same as 2 − 7, and 12 ÷ 3 is not the same as 3 ÷ 12.

NUMBER NAMES

You'll already have noticed that numbers are given a lot of impressive-sounding names. Amaze your math teacher by knowing your "integers" from your "irrationals." Read on to find out more.

What Do You Mean, Integer?

So far, this book has looked at whole numbers—starting at 0 and running along a number line with 1, 2, 3, and so on. Whole numbers, including "negative" numbers—those to the left of 0 on a number line—are known as integers. Numbers to the right of 0 on the number line are known as positive numbers.

Number Shapes

You've probably already heard of "square" numbers. These are any numbers that can be made by multiplying an integer by itself, such as 4, which is 2 x 2, or 9, which is 3 x 3. Did you know that there are also "triangular" numbers, though? These are numbers in the series 1, 3, 6, 10, and so on (see page 23).

Irrational Numbers

Numbers are "rational" if they can be obtained by dividing one integer by another, so ½, 8, and 4⅔ can be written as 1 ÷ 2, 64 ÷ 8, and 14 ÷ 3, respectively, so they are all rational. "Irrational" numbers, such as the square root of 2 (see page 25), shown as √2, and the square root of 47 (√47), cannot be written like this.*

---

*A square root is a number that, multiplied by itself, gives a particular number. For example, 4 is the square root of 16 because 4 x 4 = 16.

UNBREAKABLE NUMBERS

Most numbers can be broken down into other smaller whole numbers, or "factors." For example, 4 can be broken down into two sets of 2.

However, some numbers cannot be broken down in this way, and can only be divided exactly by themselves or by 1. For example, there are no two numbers that 13 can be divided by, except for 1 and 13. Unbreakable numbers like these are called prime numbers. The first few are 2, 3, 5, 7, 11, and 13, but there are other, much higher prime numbers that, surprisingly, can still only be divided by themselves or by 1.

Did You Know?

Mathematicians love prime numbers, and although they are easy to understand, they are very mysterious. This is because they don't fit into a regular pattern, so there is no way to find them except by trial and error.

Some mathematicians compete to see who can find the next highest prime number. The most recent one to be found has more than 12.9 million digits. It's so long that it would take more than two months to write out by hand!

NUMBERS IN NATURE

One way of describing numbers is by saying they are "odd": 1, 3, 5, and so on. Or "even": 2, 4, 6, and so on. Two odd numbers or two even numbers have what is called the same "parity"—if you have one odd and one even number, they have different parity. One of the strange things about parity is that it's not a human invention—it turns up in nature, too.

Oddly Even

The majority of animals have an even number of limbs. Humans are bipeds, which means they have two legs. Horses and dogs are quadrupeds, which means they have four legs, whereas insects have six legs and spiders have eight.

A millipede, meaning "thousand-legged" in Greek, only has around 300 legs in reality, but it is still an even number.

I LIKE THEM—I'LL TAKE 150 PAIRS.

Nature is so organized that even DNA—the blueprint for how living things develop—is usually arranged in pairs of an even number of chromosomes. Humans generally have 46 chromosomes arranged in 23 pairs.

How Odd

In the world of fruit and vegetables, different rules apply, and many of the numbers that relate to plants are odd ones. For example, lots of flowers have five petals, and fruits, such as apples, are based on a five-pointed shape, as you can see if you cut one in half to separate the top and bottom.

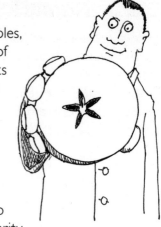

However, not all the numbers to do with plants have the same parity, but they do have something in common. Pineapples and pinecones have spikes arranged in two sets of spirals, one going clockwise and one counterclockwise.

Depending on how big the pineapple is, there will be 5 and

8 spirals, 8 and 13 spirals, or 13 and 21 spirals. These are known as Fibonacci numbers (see page 33 for more on these).

Numbers also count in the world of crystals. All sorts of things—from salt to snowflakes—are made up of crystals with regular shapes that have a fixed number of sides.

A DEADLY NUMBER

Pythagoras was a Greek mathematician who was born around 580 B.C. He hated beans and loved triangles and is considered by many to be one of the most important mathematicians ever. Little else is known about Pythagoras, except that a theorem is named after him, even though he didn't actually come up with the idea for it himself.

Lovely Triangles

Much of Pythagoras's work revolved around triangles. He was fascinated by triangular numbers, such as 3, 6, and 10, which can be arranged in a triangular pattern, like this:

Pythagoras's Theorem

Pythagoras is best known in relation to a theorem about triangles. It is a theorem that only applies to right triangles, which have one horizontal line and one vertical line. The theorem says that if you square the lengths of the two short sides and add them together, the result is the square of the length of the longest side—the hypotenuse.

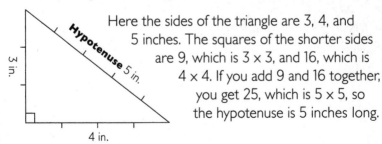

Here the sides of the triangle are 3, 4, and 5 inches. The squares of the shorter sides are 9, which is 3 × 3, and 16, which is 4 × 4. If you add 9 and 16 together, you get 25, which is 5 × 5, so the hypotenuse is 5 inches long.

23

This is very convenient, because it gives you a way to work out the length of one side of a right triangle, as long as you know the lengths of the other two sides. There's even a handy formula to show your calculations: $a^2 + b^2 = c^2$.

No Beans, Please

Although the theorem has his name on it, it is thought that Pythagoras actually heard about it in Egypt. The Egyptians and the Babylonians had been using the principle for centuries. However, he did produce some amazing theories himself. These include the basics of the science of sound, and the ideas that the Earth spins and that "everything is number."

Pythagoras and his followers were so secretive that many of their practices and ideas are unknown. Nevertheless, historians do know that as well as not eating beans, Pythagoreans would not sit on pots of a certain size and didn't allow swallows to nest under their roofs. Unfortunately, no one knows the reason why.

The Terrible Twos

Pythagoras and his followers were known to love simple things that fitted with rules. They were also convinced that all numbers could be expressed as simple ratios—for instance, one quarter is the ratio of 1:4, which is written as ¼. Then something was discovered that upset this.

Take a look at this triangle:
It looks perfectly innocent, doesn't it?

Hypotenuse

1 in.

1 in.

However, if you use the formula from the theorem to work out the length of the hypotenuse, things go a bit wrong.

This is because 1 x 1 is just 1. Add the two sides together to get the square of the hypotenuse and you get 2. There's no simple way of showing the number that, when multiplied by itself, gives the answer 2. This number, known as the square root, or √2, cannot be written as a ratio—it is an endless number that begins 1.4142 and goes on forever.

BUT, I ONLY SAID 1.41421356Z ...

This really irritated Pythagoreans, who tried to keep these endless, or irrational, numbers a secret. When one of their followers, Hippasus, kept talking about them, he was thrown to his death from a cliff!

Superstitious Stuff

The Pythagoreans aren't the only ones who dislike certain numbers. Over the centuries, particular numbers have taken on different significance around the world. For example, in China the number 8 is considered very lucky. In other countries it's the number 7. Some people believe the number 13 to be very unlucky. In fact, there's even a name for the fear of the number 13: triskaidekaphobia.

MICE ARE OKAY, BUT NOT WHEN THERE ARE THIRTEEN OF THEM.

25

How Puzzling

Number puzzles known as magic squares are believed to have originated in China. Over the centuries, some people believed that they could be used to make predictions, which isn't very likely, but they do make a good puzzle!

In a magic square, each row, column, and diagonal adds up to a "constant," which means the same number.

In this one, every row, column, and diagonal adds up to 15:

FAIR SHARES

Imagine it's your birthday and you've got seven friends over for dinner. How do you cut your cake into eight equal portions to make sure that everyone gets a fair share? It's easier than it seems, if you know how....

Nice Slices

In math, shares are called fractions, so to share your birthday cake fairly between eight people, it needs to be cut into eight equal slices. Another way of saying this would be that each slice is one-eighth of the cake. As a fraction, it would be written like this: ⅛. The top number is called the numerator; the bottom number is called the denominator.

Cutting the Cake

Here's how to make fractions work with your cake:

1. First, cut the cake in half (½) to get two equal pieces.

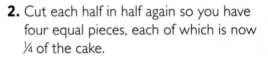

2. Cut each half in half again so you have four equal pieces, each of which is now ¼ of the cake.

3. Finally, cut each quarter in half again so that you have eight pieces. Each piece is now ⅛ of the cake— enough for you and your seven friends to dig in without arguments.

You may have noticed something else. Did you see how the fractions work together? If you halve a half, you get a quarter: $\frac{1}{2} \times \frac{1}{2} = \frac{1}{4}$, and if you halve a quarter, you get an eighth: $\frac{1}{2} \times \frac{1}{4} = \frac{1}{8}$.

Top-Heavy Fractions

For most fractions, the numerator is smaller than the denominator. These are known as proper fractions.

Occasionally, you might see a more unusual fraction, like this: $\frac{11}{2}$. This is an "improper fraction" because the numerator is bigger than the denominator. The fraction $\frac{11}{2}$ is another way of saying there are 11 halves.

Imagine these are halves of cake—if they were paired up to make whole cakes, you would have five and a half cakes, or $5\frac{1}{2}$. This is what is known as a mixed number, because it's a whole number followed by a fraction.

Perfect Percentages

What if you could divide your cake into 100 slices instead of just 8? If you wanted a quarter of the cake, you would have to take 25 slices, or $^{25}/_{100}$. This can also be written as 25 "percent," or %, which means "25 out of 100"—25% is really just another way of saying $\frac{1}{4}$.

Demystifying Decimals

Another way to think of fractions is as a division calculation. If you use the cake example, $\frac{1}{4}$ tells you to "divide 1 cake into 4 pieces." If you tap $1 \div 4$ into a calculator, the answer would be 0.25. This is what is known as a decimal fraction, or more commonly, a decimal.

If you put $1 \div 3$ into a calculator, the answer will be 0.3333333..., which goes on forever. This is called a repeating decimal. To save space, they are written with a line over the recurring part, so $\frac{1}{3}$ is written as $0.\overline{3}$.

How About Rounding?

Sometimes a number will have lots of digits after the point. For example, the number 0.4567 has four—this is called working to four decimal places. These numbers can be shortened by "rounding" up or down. If you want to work to two decimal places instead, you would look at the digit in the third decimal place—if this digit is equal to or higher than 5, as with 0.4567, you would round up to 0.46. If it is less than 5, as with the number 0.6543, you would round the number down to 0.65.

MONEY, MONEY, MONEY

In ancient times, people didn't use money; they swapped or bartered things, such as cattle. Let's face it, though, cattle aren't that easy to transport, so an alternative was needed.

In the early days of money, people used natural objects that were fairly rare and small enough to be carried easily— these could be anything from the shells of sea snails to feathers. Metal coins started to be used around 700 B.C. in Lydia, which is now part of Turkey, and in China. Even so, if people were traveling long distances or buying and selling huge sums, they still had to carry large, heavy bags of coins. In twelfth-century China, this problem was solved with the invention of paper money.

Fantastic Plastic

That's not the end of the story, though. Nowadays most of the world's money is stored on computers and is transferred electronically between banks, stores, and by people using credit and debit cards.

How Interesting

You might already have your own bank account, and maybe your savings are already earning interest, but how do you tell your interest from inflation—and what do "interest" and "inflation" mean?

The idea of interest is very simple. You put some money in a bank account, and the bank uses it to earn money for itself. In return, you get some extra money called interest. If you borrow money from a bank, you are charged interest on it, meaning you have to pay back more than you borrow.

The amount of interest you earn or pay depends on three things—the amount of money paid into or borrowed from the bank, the "rate" of interest the bank offers, and the length of time the money is in the account or borrowed for. Say the bank offers a rate of 10% interest a year on savings, and $200 is paid in and left there for a year. After that year, it will have made 10% of $200, which is $20. This means that at the end of the year, there will be $220.

Up and Up

Almost everything gets more expensive over time. Like interest, "inflation" is described by a percentage—5% is typical over a year. So if a chocolate bar costs $1 today, it may well cost $1.05 this time next year.

Luckily, most people's wages go up at about the same rate as inflation, so they can usually keep pace with price increases.

ONE AFTER ANOTHER

In math an ordered row of numbers is called a sequence. Sequences usually have a pattern, or a rule, that allows you to predict the next number. Counting puts numbers into about the simplest sequence there is. In the sequence below, 1 is added to the previous number:

$$1, 2, 3, 4 \ldots$$

When numbers in a sequence are added together, the result is called a series:

$$\tfrac{1}{2} + \tfrac{1}{4} + \tfrac{1}{8} + \tfrac{1}{16} \ldots$$

This series is making its way toward the number 1. Adding the first two terms gives $\tfrac{3}{4}$, adding the next, $\tfrac{1}{8}$, to $\tfrac{3}{4}$ gives $\tfrac{7}{8}$. Then $\tfrac{7}{8} + \tfrac{1}{16} = \tfrac{15}{16}$, and so on.

A Rat's Life

Series and sequences occur in the real world, too. For instance, the total number of rats after a certain number of generations is given by a series. Rats usually have about ten babies in each litter. The numbers in each generation are given by multiplying the number in the last generation by 10:

$$1, 10, 100, 1,000 \ldots$$

The total number of rats after the great-grandkids are born is:

$$1 + 10 + 100 + 1,000 = 1,111$$

The number in each generation can be obtained just by adding a 0 to the last number, so in the tenth rat generation there would be

1,000,000,000 rats. Scary, eh? This sort of thing leads to some big numbers fast, which goes to show that you have to be careful in applying math to the world—in this case, the assumption is that each baby rat survives to bear a litter of little rats, but in reality, not all of them will make it and breed, thank goodness.

The Fibonacci Sequence

Another sequence that can be applied in nature is known as the Fibonacci sequence. It goes like this:

1, 1, 2, 3, 5, 8, 13, 21, 34 ...

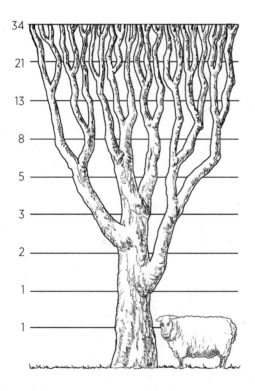

There is a simple rule to make this sequence. Can you see what it is? (The answer is at the bottom of the page.)

Answer: Fibonacci numbers: Add the most recent number to the one before it to get the next number in the sequence.

This sequence turns up in some odd places. The pineapples on page 22 have two consecutive Fibonacci numbers in their spirals, and they are also found in the way that tree branches grow, the arrangements of seeds in sunflower heads, and the number of honeybees in successive generations.

Patterns in Pennies

You might even have Fibonacci numbers in your pocket.

If you have any 5-cent and 10-cent coins, how many ways can you make 15 cents? You could have:

5 cent and 10 cent

10 cent and 5 cent

Three 5 cents

So there are three ways to make 15 cents. What about 20 cents? You could have:

5 cent, 5 cent,
5 cent, and 5 cent

10 cent, 5 cent,
and 5 cent

5 cent, 10 cent,
and 5 cent

5 cent, 5 cent,
and 10 cent

10 cent and 10 cent

So there are five ways to make 20 cents with 5- and 10-cent coins. And—you guessed it—there are eight ways to get 25 cents, 13 ways to get 30 cents, and 21 ways to get 35 cents. How cool is that?

SIMPLY ENORMOUS

The universe is utterly enormous. Just a single drop of water contains about ten billion trillion atoms, or 10,000,000,000,000,000,000,000. If you think that's a lot, the number of atoms in the ocean is about a trillion trillion times more than this. Not surprisingly, mathematicians have come up with a far easier way of showing these huge numbers. For example, a trillion trillion is written as 10^{24}, which means "10 to the power of 24," or "10 multiplied by itself 24 times," or 1 followed by 24 zeros. This is called scientific notation.

The Complex Googolplex

Some very large numbers have special names. For example, 10^{100} is a googol—the name suggested by nine-year-old Milton Sirotta in 1938. That's not all—there's also the googolplex, which is 10 to the power of a googol, or $10^{10^{100}}$. And did you know that the search engine Google is based on an alteration of this name?

That's a Lot of Sand

The first person to study huge numbers was Archimedes, in ancient Greece. In his work *The Sand Reckoner,* he estimated

the size of the universe and how many grains of sand it would take to fill it. He took a myriad, the largest number in use at the time—10,000—and, by multiplying it by itself again and again, calculated that the whole universe contained 8×10^{63} grains of sand. Since the universe is quite possibly infinite, he was way off, but it was still impressive work for the time.

BEYOND INFINITY

Imagine a number line (like the ones on pages 17 and 18); then imagine it extends forever in both directions. This is infinity. There is no largest number, because you can always add one more to make a bigger number.

In many ways, infinity is similar to zero, which cannot be divided. For example, half of infinity is infinity, and even a trillionth of infinity is still always infinity.

Infinitely Complicated

So does infinity mean anything? Well, it might, because you might live in an infinite universe. If that's true, there may be an infinite number of stars. If just one in a trillion stars had a planet similar to Earth orbiting it, that would mean there would be an infinite number of planets like Earth, because a trillionth of infinity is infinity. If there are an infinite number of planets like Earth, there must be an infinite number of planets with humans living on them, and an infinite number of those humans must have your name, your age, your parents—there must be an infinite number of people just like you. It's enough to make your head spin!

I TOLD YOU WE WENT THE WRONG WAY—WE PASSED OUR EARTH TWO GALAXIES BACK.

STARTLING SHAPES

ALL KINDS OF SHAPES

A lot of mathematics is about simplifying things, from human behavior to the motions of the planets. Geometry—the mathematics of shapes—is no exception.

Natural Wonders

Many things in nature, such as trees or humans, for example, have complicated, irregular shapes that take a lot of describing.

Regular shapes don't occur naturally very often, but there are some, such as the disk of the sun or the hexagonal shapes of a honeycomb in a beehive. In a honeycomb, each cell is hexagonal. This maximizes the space in each one, while keeping the gaps between them to a minimum.

If circles were used, more wax would be needed to fill the gaps between the cells. Isn't nature clever?

Where's Polly Gone?

Simple, regular shapes can be divided into shapes made of straight lines, such as squares, and shapes made of curves, such as circles. Straight-line shapes are called polygons.

Those with sides of equal length are called regular polygons. These include:

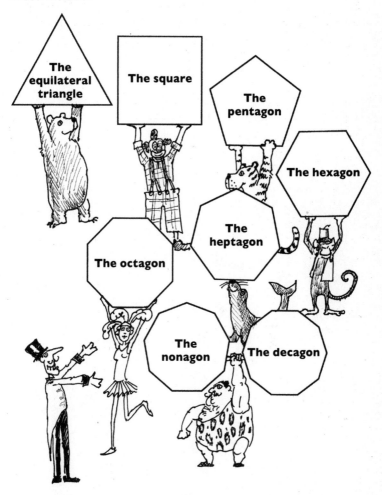

After the square, the names of polygons settle down into a regular pattern. Each shape has the Greek word for the corresponding number of sides, followed by "-gon"—the Greek word for angle.

TERRIFIC TRIANGLES

Other than circles, triangles are probably the shapes that mathematicians love best. So much so that the study of triangles has its own name: trigonometry.

The Strongest Shape

The triangle is one of the simplest shapes because it has the smallest number of straight lines—and math is all about simplifying. It is also one of the most useful shapes for building structures. In wet climates, many roofs are "pitched," or triangular, so that the rain can run off easily. Cranes, bridges, and many other structures are made of triangles, too. The reason is simple—it's the strongest shape.

Triangle Types

Triangles can be grouped into three kinds. These are:

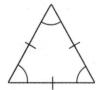

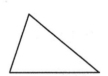

Equilateral triangle
All three sides and all three angles are equal.

Isosceles triangle
Two sides and two angles are equal.

Scalene triangle
No sides and no angles are equal.

The three angles of the corners of a triangle always add up to 180°, no matter what type it is.

Did You Know?

Angles of less than 90° are acute angles. Angles between 90° and 180° are known as obtuse angles. And those between 180° and 360° are reflex angles.

Similar Triangles

If you wanted to find the height of something very tall, you'd need a long ladder and a huge tape measure, right?

Wrong. All you need to do is find a long, straight stick and something to measure—a nice, tall tree, for example. Push the stick into the ground a little way from the tree. Next, find a position on the ground where the top of the tree lines up with the top of the stick from your point of view. This will make two triangles—one from your eye to the stick, and the one from your eye to the tree. They both have the same angles, even though they are different sizes. These are known as similar triangles.

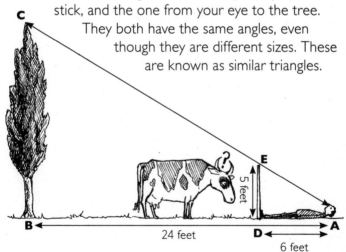

The lengths of each side of similar triangles are also in proportion. With a few simple measurements, you can use this to work out the tree's height compared with the stick.

I USED TO KNOW THAT: MATH

To do this, you need to measure the distance from A to D and the distance from A to B. Then divide the length of AB by the length of AD to get the ratio, which would be 4 feet in the example on the previous page. By multiplying the height of the stick by this ratio, you can find out the height of the tree.

In this case, $4 \times 5 = 20$, so the tree must be 20 feet tall—and all without going up a wobbly ladder!

Out of This World

Some distances are even trickier to measure directly—the distance to the moon, for example—but triangles can still be used to calculate them. All you need to do is measure the length of one side of an imaginary triangle, between points A and B, to get a baseline. Then measure the angles made by the imaginary lines to the moon at A and B. Since the Earth and the moon are both moving, this must be done at the same time when the moon is between the two. Once you know those angles, the length of lines AC and BC can be worked out, which also gives you the distance to the moon. This process is called triangulation.

Moon

C

Distance to the moon

A   B

Reach for the Stars

If you tried to triangulate a star using the same baseline as above, it would be too narrow because stars are so distant. There's a clever trick to increase the triangle's size, though.

By measuring angle A first, then angle B six months later—when Earth is directly opposite in its orbit—the distance between points A and B will be double Earth's distance to the sun. Using this interstellar-size triangle, you can work out the distance to the star. Simple!

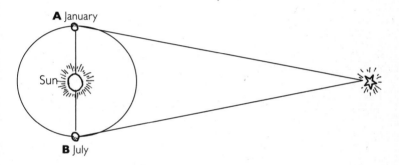

ALL ABOUT AREA

Simple, regular polygons, such as squares or rectangles, are easy when it comes to calculating area—the amount of space they take up. Just multiply the length by the width. For a square measuring 4 in. by 4 in., you simply multiply 4 x 4. The answer is 16, so the square has an area of 16 in.^{2} A rectangle that is 4 in. by 2 in. has an area of 8 in.^{2}

Equally, if you want to know the distance around the edge of your square or rectangle—its perimeter—you just combine the lengths of all four sides. How would you measure a circle, though? There are no sides for you to multiply, and sometimes it would be really useful to be able to figure out a circle's area or its circumference, which is the special name for the perimeter of a circle.

As Easy As Pi

The answer is a lot simpler than you may think. For centuries, people knew that for any circle, the circumference is just over three times the diameter—the distance across the middle of a circle. This ratio stays the same whatever the size of the circle.

As time went on, people improved the calculation, and in fact, the circumference of a circle is more like 3.141592654 times the diameter. This number is called pi, which can also be written with the symbol π and is often abbreviated to 3.142. It is another irrational number (see page 19) and actually goes on forever.

(Circumference = 3.142 × D)

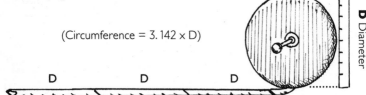

D Diameter

D   D   D

The Area of a Circle

The way to calculate the area of a circle involves pi again, but this time working with the radius, which is half the diameter. You just need a special formula: πr^2, or "pi R squared," which means the area of a circle is equal to pi × (radius × radius). For example, if a circle had a radius of 4 inches, the radius squared is 16. If you multiply π by this number, you get the area of the circle: $\pi \times 16 = 50.272$ in.^{2}

To look at the area of a circle in a more visual way, try cutting up a circle of paper, like this:

1. Fold your circle in half, then in half again crosswise, dividing it into quarters, as shown.

2. Now fold the creases so that they meet and press down. Repeat at a right angle to this fold. Your circle will now be divided into eight equal segments.

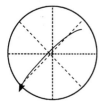

3. Cut up the segments and lay four of them out in a row.

4. Slot the other four segments in to make a slightly bumpy, cockeyed rectangle called a parallelogram, with the same area as the circle.

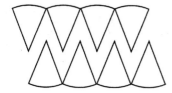

The long side of this parallelogram measures the same as π × r; the height is the same as r. To put it another way: π × r × r, or r^2. The more equal segments you cut your circle into, the more it will look like a rectangle.

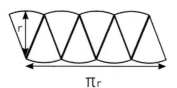

πr

47

INTO THE THIRD DIMENSION

Flat shapes, such as triangles, squares, and rectangles, can be combined in various ways to make 3-dimensional shapes, or polyhedrons. For example, a cube is made with six squares; a cuboid can be made up of three pairs of rectangles.

What's Inside?

The amount of space these shapes contain is called the volume. You can easily find the volume of a cube or cuboid by multiplying its length, width, and height together.

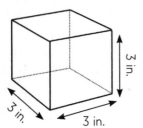

For example, to find the volume of a cube with edges that are 3 inches long, you just multiply 3 by itself and then by itself again: 3 x 3 x 3. So 3 x 3 is 9, and 9 x 3 is 27, which means the cube has a volume of 27 in.^{3}, or 27 cubic inches.

A cuboid with sides of 2, 3, and 5 inches will have a volume of 2 x 3 x 5, which is 30 in.^{3}

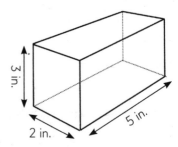

Simple Cylinders

People usually think of a cylinder as being shaped like a can of soup. However, the faces can be any shape, as long as

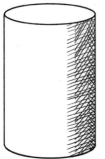

the cylinder has the same cross section along its height.

If the faces at the ends of the cylinder are parallel, the volume will always be the area of the cross section multiplied by the height.

To find the volume of a circular cylinder, you need the formula $\pi r^2 h$—that's the formula for finding the area of a circle

(page 46) multiplied by height, or "h." So a circular cylinder that is 5 inches tall with faces that have a radius of 3 inches would have a volume of $\pi \times 3 \times 3 \times 5$, which is 141.372 in.^{3}

Spherically Speaking

Humans might like shapes such as cuboids, but nature prefers the sphere—the most regular shape possible—from stars, planets, and many moons to bubbles, eyeballs, and atoms. This is because a sphere uses the least possible surface area to enclose the maximum volume.

To find the volume of a sphere, you need to know the radius— the measurement from the

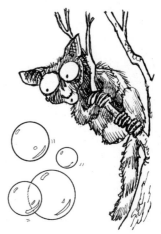

exact center of the sphere to its edge. To find the volume, you must cube the radius, multiply this number by 4π, and then divide it by 3. This formula is usually written as $\frac{4}{3}\pi r^3$. For example, a sphere with a radius of 3 inches will have a volume $27 \times \frac{4}{3} \times \pi$ or 36π, which is 113.097 in.^{3}

Shapes and Space

There are just five 3-dimensional shapes that have equal sides, equal angles, and equal faces. These are known as regular polyhedrons, or Platonic solids. The ancient Greeks were very keen on geometrical shapes—in fact, they were so keen on Platonic solids, they believed that they could be used to explain how the universe was constructed.

The ancient Greeks believed that everything on Earth was made from four elements: earth, fire, water, and air. They also thought that each one was made up of atoms shaped like one of the Platonic solids, like this:

Tetrahedrons
(made up of four equilateral triangles): fire

Octahedrons
(made up of eight equilateral triangles): air

Dodecahedrons
(made up of 12 pentagons): stars and other planets

Cubes
(made up of six squares): earth

Icosahedrons
(made up of 20 equilateral triangles): water

Terrific Topology

To study shapes that are more complex than regular polyhedrons, "topology" is used.

Topology studies what happens to shapes when you squash, stretch, or twist them. When shapes can be stretched or twisted to make another shape, they can be described as topologically equivalent. For example, a doughnut and a needle can be described as topologically equivalent, because they both have a hole going through them. Neither can be described as topologically equivalent to a glass, however, as a glass does not have a hole going through it.

Mind-Bending Möbius

Topologists have made some rather weird discoveries about shapes. For instance, 3-dimensional objects are thought of as having an inside and an outside, but in 1858 the German mathematician August Möbius created something that had just one continuous surface. It is known as the Möbius strip.

You can make your own Möbius strip quite easily. Take a strip of paper, twist it once, and stick the ends together so that it looks like this:

Take a pen and draw a line along the middle of the strip without lifting it off the paper. Your pen will cover every bit of the strip, then come back around to meet your starting point, proving that the Möbius strip has just one surface.

Baffling Bottles

A man named Felix Klein came up with an even more incredible idea. The bottle named after him is a shape with just one side—a continuous one-sided cylinder, in a way. In fact, a true Klein bottle would be 4-dimensional, as it needs to pass through itself without making a hole, which is impossible in three dimensions.

INNER SPACE

The Klein bottle on the previous page is an example of a 4-dimensional model—well, a 3-dimensional model of a 4-dimensional shape, but what does that actually mean?

A shape with no dimensions looks like this....

... Well, like nothing, really. Zero-dimensional shapes occupy no space, so there's nothing to see.

The First Three Dimensions

One-dimensional shapes can't really be seen either, as they are just lines with no thickness. However, in mathematics it's normal to imagine that lines are just thick enough to allow them to be drawn. So here is a one-dimensional shape:

By extending it sideways, you can make a 2-dimensional shape. Here it is a diamond, or rhombus:

If you add some vertical lines and another rhombus, it becomes a representation of a 3-dimensional shape—a rhomboid.

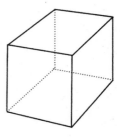

What about the fourth dimension— how would that look?

53

Into the Fourth Dimension!

To look at the fourth dimension and beyond, imagine a square. A 2-dimensional square is made with 4 lines— a 3-dimensional cube is made with 12 lines. The next square shape—in the fourth dimension—has 32 lines and is known as a tesseract, or hypercube.

The best way to think of a tesseract is to imagine two 3-dimensional cubes attached to each other, so that every face of the cube becomes a 3-dimensional shape, like this:

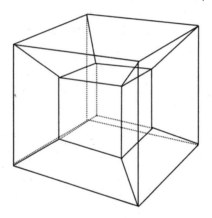

Did You Know?

Time is also a dimension. You travel from morning to night, just as you travel to and from school. Although, of course, reversing, turning, and stopping in time isn't possible—yet!

Space-Time

Time and space are always linked—they are known as space-time. The best-known scientist to work with space-time was Albert Einstein. He discovered that when things travel very quickly, they change shape.

Imagine that there was a big bomb traveling through space. Einstein proved that if it was going fast enough, it would change shape, like this:

How Special!

Another part of Einstein's space-time theories said that time passes at different rates depending on how the person measuring it is moving. If you were unfortunate enough to be on the bomb, you might measure the time for its fuse to burn down as a minute. If someone on the ground watched you whiz past and timed the fuse, too, he might measure it as taking two minutes. To look at it another way, imagine a spaceship traveling at nearly the speed of light. If the astronauts on board traveled for what they measured as a year, they would find that far longer had passed on Earth when they returned to it—perhaps thousands of years.

An Extra Dimension or Three

Does the world really have just four dimensions? Of course not—that would be silly. In reality, scientists think it has 11!

"M-theory"—the idea of 11 dimensions—suggests that 10 dimensions are of space and 1 is time.

How come no one has ever noticed all these extra dimensions, though? The answer is that they are scrunched up too small to see, or compactified. Perhaps this is just as well—a 10-dimensional equivalent of a cube, called a dekeract, is made up of 5,120 lines—imagine that!

Did You Know?

Einstein's theories also say that the speed of light is the fastest thing possible. Until very recently, everyone was still certain that was true. However, scientists in Switzerland at the European Organization for Nuclear Research (also known as CERN) discovered that it might be possible for tiny particles, called neutrinos, to move faster.

PERFECT SYMMETRY

Many shapes, and other things, too, have symmetry, which means that if you were to cut them in half from top to bottom, the two halves would be similar, if not identical.

Even people have symmetry. If you were to draw a line down the middle of your body, from the top of your head to your feet, the two halves would be very similar. The line you drew down yourself is called an axis of symmetry.

How Many Axes?

Many regular shapes have several axes of symmetry. This square, for instance, has four axes of symmetry. Can you work out how many axes of symmetry a circle has? The answer is at the bottom of the page.

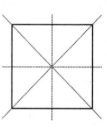

MIRROR, MIRROR, ON THE WALL, WHO IS THE MOST SYMMETRICAL OF THEM ALL?

Mirror, Mirror

Take a mirror and hold it at a 90° angle against a photo of your face along the length of the middle, so that the left side of your face is reflected in the mirror. Turn the mirror and repeat it with the right side of your face. Do the images look similar or different? The more similar the images are, the more symmetrical your face is.

Answer: A circle has an infinite number of axes of symmetry. **57**

MATHEMATICAL STRUCTURES

Man-made structures come in many shapes and sizes. Some are pleasing to the eye, and others seem downright unattractive. Part of the reason why some look more appealing can be summed up mathematically.

The Golden Ratio

The United Nations building in New York and the Parthenon in Athens both have the same ratio of length to height: 1:1.618. This is known as the golden ratio. It can also be represented by the Greek letter phi, or "Φ," (not to be confused with pi!). As well as cropping up in architecture, it also pops up in art. It is often used for the shape of photos, books, and postcards.

The golden ratio has some interesting mathematical properties, too. If you take a rectangle with a golden ratio like the one below, divide it into a square and another rectangle, the smaller rectangle will also have the golden ratio. This can be repeated again and again.

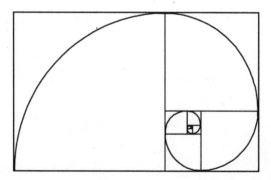

What's more, if you draw a curve between opposite corners of each square, you will create a spiral.

THE ART OF MATH

The farther away things are from you, the smaller they appear to be. The basic rule is that if something is twice as far away, it appears to be half the width and height. For artists, learning how to draw or paint things in perspective—so that they look the right size—is an important skill.

The Point of Vanishing Points

Imagine standing in the middle of a straight, flat road. The sides of the road would seem to get closer together the farther away they were until they looked as though they joined in a point. In art, this is called the vanishing point.

Putting It in Perspective

If you look at any painting created before the fifteenth century, you'll probably find it looks odd—artists hadn't really mastered the art of perspective yet.

Vanishing point

Horizon line

If you are drawing a picture that has lots of parallel lines—a straight, flat road full of lots of houses, for instance—all the objects will disappear to the same vanishing point. This means you can draw in guidelines to help you check that your perspective looks right.

MUSIC BY NUMBERS

Math is very useful for music, too. In fact, it's numbers that make the difference between music and a terrible din.

Here's what happens when you strum or pluck the string of an instrument, like a guitar:

When you strum a guitar, for instance, the wobbling string makes this kind of shape.

If you press your finger halfway along its length, its shape changes to this.

When you do this, the sound that the string produces rises in pitch, so that it sounds higher, but it still sounds similar. In music these two notes are connected with the same letter name—C, for example. These two C notes are described as an "octave" apart—an interval of eight notes: C, D, E, F, G, A, B, and C. An interval is the distance between two notes.

The sounds that the string makes are actually waves, a little like the ripples you see when a pebble is dropped in a pond, except that they are traveling in all directions through the air.

Perfect Harmony

Notes that are in harmony sound good together because they make simple ratios. In notes an octave apart, the ratio is 2:1. The higher-pitched note has a wavelength that is exactly half as long as the other, which makes sense if you remember it is produced by halving the length of the string.

When two notes an octave apart are played together, they sound good. Other combinations sound good, too. For example, an interval of a fifth is a pair of notes where one has a wavelength one-and-a-half times as long as the other. Notes that sound unpleasant together have different wavelengths, which clash because they don't make simple ratios—just like the sound they produce.

61

Pythagoras and Planets

Your old friend Pythagoras (see pages 23–25) first noticed that the pleasant or harsh sounds made by different combinations of notes involved ratios.

He was so set on his ratios idea that he thought the entire universe could be explained with numbers. He and his followers believed that the distances between planets were simple ratios and that they made harmonious sounds as they moved. This idea was called the harmony of the spheres and was popular for many centuries.

 ___**A MATTER OF DEGREES**___

According to many people—especially everyone who has to go up it—Baldwin Street in Dunedin, New Zealand, is the steepest street in the world. This has been subject to much discussion, largely because there are so many ways to measure its "gradient"—or steepness.

How Is It Measured?

Steepness can be measured using a ratio. Baldwin Street's ratio is 1:2.86. This means for every 2.86 m (9.38 feet) traveled horizontally, the street rises by 1 m (3.28 feet).

Percentages are sometimes used to measure a gradient instead. Baldwin Street's gradient is 35%, which means that for every meter traveled, the street rises by 35 cm (14 inches).

The gradient can also be expressed as an angle in degrees, known as an angle of elevation—the amount that it rises from an imaginary horizontal position on the ground. Baldwin Street's angle of elevation is 19.3°.

SCALING THE HEIGHTS

Old science-fiction films were full of giant killer insects and other monsters, but why are they so small in real life? The answer is mathematical, of course.

An Ant Experiment

Insects, such as ants, don't have a skeleton like you. Instead, their hard outer shells, or exoskeletons, support their bodies. An insect also breathes through its exoskeleton, but there is a limit to how much air it can absorb. That amount depends on the insect's surface area. However, the amount of air an insect needs depends on its volume. The problem is that volume and area don't get bigger at the same rate. Look at the table below to see what would happen if an ant became 10 inches long, or even 100 inches long:

Ant Size	Length (in.)	Surface Area (in.^{2})	Volume (in.^{3})
Small	1	1	1
Large	10	100	1,000
Giant	100	10,000	1,000,000
Monster	1,000	1,000,000	1,000,000,000

The small ant's volume and area are fine—it can absorb as much air as it needs through its exoskeleton. However, as the size of the ant increases, the volume increases far more than the surface area. By the time the ant is monster-size, its volume is a thousand times its surface area. This means that the ant couldn't possibly have a large enough surface area to provide the oxygen its body would need. (See pages 49–50 for more on volume.)

Sizing Things Up

Look at the picture of an elephant shrew and an elephant. Even though they look the same size here, you know automatically that the shrew is far smaller than the elephant. What if an alien mathematician looked at them, though?

Surprisingly, the alien would probably know that, too. This is because of the shrew's tiny legs. They are too thin in proportion to its body compared with the elephant's legs.

Don't Fall!

When something falls—an elephant, say—the air pushes back up against it. This force, called resistance, depends on the falling object's area, but the force downward depends on its weight. The object's weight depends on its volume, which also depends on how dense* it is. The smaller the object, the more air resistance there is per pound, which means it is slowed more than a larger object would be.

This means that if you were unlucky enough to fall from a height of 13 feet, you'd have to go to the hospital. A spider would hardly notice, while an elephant ... well, don't worry about giant ants, but do watch out for falling elephants!

---

*Density relates to mass—an object the size of a Ping-Pong ball, but made of iron, would have more density than an object the same size made of cheese.

AMAZING MEASUREMENTS

MEASURING MONARCHS

Not so long ago, people didn't calculate length and height using rulers and tape measures. In fact, a very different approach was taken. Some measurements were based on body parts, which is why the height of horses is still measured in hands and the height of people is given in feet and inches.

Handy Measurements

In ancient Egypt, things were measured in something called a cubit, which was the length of the current pharaoh's forearm, plus a hand width. The length was carved into a granite block, and wooden or stone copies were given to builders. This worked fine, but only for as long as the pharaoh lived. Lengths were likely to change as pharaohs came and went, which was probably quite annoying if you were halfway through building something.

There were also a dozen different lengths—also called cubits—known to have been used elsewhere in the ancient world, which must have been very confusing.

Inching Forward

Over the next few centuries, many measuring systems were used around the world, and these could vary within countries. It wasn't until the thirteenth century that people tried to standardize measurements, when it was ruled that an inch was the length of three barleycorns. Over time the imperial system was adopted: the inch, foot (12 inches), yard (3 feet), and mile (1,760 yards). The United States still uses this, but in Europe it was gradually replaced by the metric system of centimeters, meters, and kilometers. But who decided all of that?

CURIOUS UNITS

In 1960 the International System of Units *(Système international d'unités,* abbreviated as SI) was formally agreed upon as a standard of measurement.

Incredible Seven

There are seven standard SI units. There's the meter for distance, the kilogram for mass, and the second for time. Then there's the ampere for electric current, the kelvin for temperature, the mole for an amount of substance, and the candela for luminous intensity, or brightness of things. These units are used as a jumping-off point for all other types of metric unit.

What's more, even though there are so many kinds of things that you might want to measure—such as loudness, cloudiness, smoothness, and sharpness—the seven SI units can measure them all, thanks to mathematics.

Really Big or Really Small?

What happens if you want to measure something huge or something tiny using SI units? After all, feet and meters are good for measuring the height of a tree, but probably not so good for measuring the size of its leaves. And if you

69

want to measure the size of a whole forest, you're likely to need something much larger.

Which Unit Is Best?

By using the seven main SI units as a starting point, you can scale each type of unit up or down to measure all sorts of things. All you do is add a different "prefix," or beginning, to the words to indicate that a larger or smaller measurement is being used. For example, the "kilo-" part of kilometer means "one thousand," the "centi-" part of centimeter means "one hundredth," and the "milli-" part of millimeter means "one thousandth."

If you wanted, you could measure distance traveled in centimeters, but it would be tricky for a motorist to know which turns to take. Instead, it's best to try to choose the most suitable measurement for the thing you are measuring.

...CONTINUE STRAIGHT FOR 100,000 CENTIMETERS AND YOU WILL REACH YOUR DESTINATION.

An exception to these rules is the way in which time is measured. There was already a pretty much internationally agreed system in place long before SI units came along, so a minute is still 60 seconds and an hour is still 60 minutes. For smaller units of time, though, there's the millisecond and even the nanosecond—that's a billionth of a second!

Red-Faced on the Red Planet

SI units have been adopted by most countries around the world. This means that when people are working with other nations to build things, everyone is working from the same set of measurements. That may not seem so important, but when no one is sure which unit of measurement is being used, the results can be chaotic, with very expensive consequences!

For instance, in 1999 NASA's *Mars Climate Orbiter*, which cost $125 million to build, had successfully completed nearly all of its journey. It was on its final approach to the Martian atmosphere when things went seriously wrong. One team was working in old-fashioned imperial units, while the other was working in metric. The result was that the orbit was miscalculated and the orbiter crashed.

PRECISELY ACCURATE

The words "accurate" and "precise" might sound as though they can be used interchangeably. In reality, they have slightly different meanings.

Right on Target!

An archer has shot five arrows into each target below:

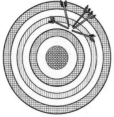 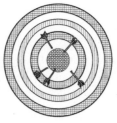 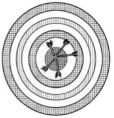

Precise, but not accurate   Accurate, but not precise   Precise and accurate.

On the first target, all the arrows are close together, so the archer has managed high precision. However, they are not near the center of the target, so his accuracy is low.

On the second target, the archer has achieved a high level of accuracy, as all the arrows are close to the center of the target. In this case, though, the arrows aren't close to each other, indicating that his level of precision is low.

As you can see, on the final target, the archer has shot the arrows both accurately *and* precisely.

"GUESSTIMATION"

Being able to estimate is a useful skill and one that most humans are good at. For example, did you know that when trying to judge whether something is at a right angle or not, most humans can spot errors of less than a degree?

The ability to estimate can also help with everyday things, such as crossing a road. By estimating the speed and distance of vehicles, you can judge whether or not you have enough time to cross a road safely.

TOO FAST!

I'M SURE THERE'S ENOUGH TIME TO...BLEURGH!

However, people don't think in terms of units naturally. Rather than thinking, "That car is moving at about 20 miles an hour," you are more likely to think in terms of, "That's too fast" or "There's enough time to cross."

Simple Samples

Estimating can save time, too. For instance, you might be asked to find out the type and number of different creatures living in the turf of a playing field. Counting them all would take a ridiculous amount of time. Instead, you could estimate the answer using a process called sampling.

To sample an area, the first thing that's needed is a quadrat—a square made of four rods joined together, each a yard long. This is placed on the playing field, then all the

73

living creatures found in that area are collected and recorded. For example, you might find 4 earthworms, 16 ants, 3 beetles, and 1 slug.

Now say that the playing field measures 100 by 70 yards. If you multiply these figures together, you can find out the playing field's area: $100 \times 70 = 7,000$ square yards.

To estimate how many creatures there are, simply multiply the numbers of the quadrat populations by 7,000 to give:

AAARGH!

28,000 earthworms,
112,000 ants,
21,000 beetles,
7,000 slugs.

Of course, this type of approach has to be used with care. If you happened to find a coin in the square, too, it's very unlikely that there would be a fortune scattered across the rest of the field.

Surprising Guesses

Making estimates can lead to some amazing facts, and even to some disturbing ones. For example, Shakespeare's dying breath in 1616 would probably have contained around 1,000,000,000,000,000,000,000 air molecules. Over the next four centuries, these molecules have spread across the world and through the atmosphere, which contains about 1,000,000,000,000,000,000,000,000,000,000,000,000,000,000,000 air molecules (that's a million, billion, billion, billion, billion).

If you divide the larger number by the smaller one, you get 1,000,000,000,000,000,000,000. This means that you can estimate that 1 in every 1,000,000,000,000,000,000,000 molecules in the air is from Shakespeare's last breath. However, this also happens to be the same number of molecules you breathe in every breath, so there's a good chance that your next breath includes a molecule from Shakespeare's last one!

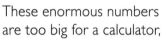

These enormous numbers are too big for a calculator, but there is a quick way to work with them. As you know from page 36, long numbers such as 1,000 can be expressed as powers of 10, so 1,000 can be written as 10^3, because it is the same as $10 \times 10 \times 10$. The number of molecules in Shakespeare's last breath can be shortened to 10^{21}. The number of molecules in the atmosphere is 10^{42}.

Handily, dividing these numbers gives the same answer as subtracting their powers of 10. So the question above can also be worked out as $10^{42} \div 10^{21}$, which is 10^{21}.

A Calculated Guess

Estimation can also be useful when using calculators and spreadsheets to work things out. This is because if you have a rough idea of how much the total will be, it's much easier to spot when a mistake has been made, so that you can check if the wrong button has been pressed.

FOOLING YOURSELF

The fact that the human brain can estimate all sorts of things means it needs to make assumptions. For example, you might spot a dot on the horizon that gets bigger until you see a tiny person. If that person gets larger, you would assume that he was approaching you and that he was about your size. Once these assumptions are made, your brain can use the laws of perspective (see page 59) to judge the distance of the person and his speed of approach.

The problem with assumptions is that they can sometimes be wrong, which can result in an optical illusion, like this:

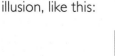

Your brain will try to make the best sense it can of what you are looking at.

At first, it might seem as though you're looking at an image of a box seen from above. However, after a while, a different perspective might be seen of a box viewed from below. Which one is correct?

Either is correct. The image of the box is an optical illusion known as a Necker Cube. The human brain tries to make sense of it when really there isn't enough information, which is why you can see two perspectives.

drop a feather and a cannonball on the moon, where there is no air, they would fall at the same rate.

By imagining how things would fall if there were no air, Galileo recognized that the laws of falling objects were very simple. He realized that they could be described with some straightforward mathematics:

> Objects fall with a steady, or "constant," acceleration. So if an object is falling at a particular velocity, after 1 second it will be falling twice as fast; after 2 seconds, 3 times as fast; after 3 seconds, 4 times as fast, and so on.

What's more:

> If an object has fallen through a particular distance in 1 second, it will have fallen through a total of 4 times that distance after 2 seconds, 9 times the distance after 3 seconds, and 16 times the distance after 4 seconds.

Have you noticed how the two sets of numbers are related? The distances—1, 4, 9, and 16—are the squares of the times—1, 2, 3, and 4. The longer an object falls, the faster it travels.

Did You Know?

You might wonder about the point of these ideas and discoveries if air resistance means they don't apply. Well, on Earth, adjustments can be made to mathematical equations to take this into account. And if you traveled into space, you'd soon find them useful—spacecraft obey these rules because there is no air to slow them down.

BLINK AND YOU'LL MISS IT

The briefest moments of time are measured in less than a trillionth of a trillionth of a trillionth of a "nanosecond," or billionth of a second. A nanosecond is so quick that it takes 100 million nanoseconds to blink just once, so it is not a measurement that is useful for everyday life.

> I MISSED THAT LAST 100 MILLION. SORRY—I BLINKED.

Only a Matter of Time

You might not spot a nanosecond easily, but time is something that you are probably very aware of. Whether it's minutes dragging by during a boring class or hours whizzing past as you enjoy an over-too-soon birthday party. What is time exactly, though, and how is it measured?

Early humans measured time passing in various ways—including the changing of the seasons, or the shape of the moon over the course of a month. Eventually, however, people wanted more accurate ways to measure time. To do this, they came up with some useful units.

Minute after Minute

You already know that there are 60 seconds in a minute, 60 minutes in an hour, 24 hours in a day, seven days in a week, and 365 days in a year. Why is it that months don't follow a pattern, though? Some have 30 days, others 31 or even 28 or 29. To find out why, you'll need to go back in time to see how it was first measured.

CLOCK THIS

One of the earliest ways to measure the passage of a day was developed in ancient Egypt around 3,500 years ago. The shadow clock, as it was called, relied on the shadow cast by the sun as it moved across the sky. It worked because a shadow's length and direction change as the sun moves.

Playing with Shadows

The shadow clock was quite simple and divided the hours of daylight into 10, plus an hour of twilight in the morning and one in the evening. It consisted of a raised bar that cast a shadow over another bar, along which "hours" were marked. In the mornings, the raised bar pointed east, so that the shadow fell on the hour bar to the west. At noon, the clock would be rotated to point in the other direction and follow the sun as it moved from east to west. What's more, a shadow clock was portable, although it had to be aligned just right each time it was moved.

However, a shadow clock could not be used to tell the time at night, and if the sky was cloudy during the day, you wouldn't have any idea if you were right on time or super late. Another problem was that the farther away from the equator you were, the more the lengths of days differed throughout the year. This would mean that in winter, your shadow clock would divide the day up into the same number of hours, but they would be shorter.

THE FACE DIVIDES THE DAY INTO 12 HOURS.

Although no one is sure exactly where or when sundials were first used, they look much more like the circular clocks and watches you see now. They were a big improvement on shadow clocks, because the hours they measured were the same length throughout the year. However, there were still problems—they had to be properly aligned, and they relied on sunlight to show the time.

Clocks for Cloudy Days

What was needed was a clock that could measure time inside or out, whatever the weather, day or night.

The ancient Egyptians experimented with water clocks. The earliest surviving example dates from around 1500 B.C. These clocks worked by allowing water to drip at a steady rate into a bowl where the "hours" were marked up the side. However, they still weren't very accurate.

Things finally improved with the invention of the mechanical clock sometime in the fourteenth century. It worked, thanks to a gradually falling weight, which turned a system of gears. These clocks *still* weren't very accurate, though.

What was really needed was something to make sure that the speed of the dropping weight was constant so that the gears turned at a constant speed, too. Galileo (see page 78) is known to have sketched some ideas for a clock that used a pendulum in 1642, but the first was made by the Dutch mathematician and astronomer Christiaan Huygens in 1656. Its swinging pendulum made this type of clock the most accurate tool for timekeeping until the twentieth century.

Getting There!

In the seventeenth century, people knew the Earth rotated 360° in a day and 15° an hour. If you sailed east for an hour, you knew that your location was 15° east of your starting point. If you were a few minutes off, your location would be wrong. Pendulum clocks, although accurate on land, wouldn't work at sea. In 1759 John Harrison's H4—an oversized but accurate pocket watch—solved this problem.

Meanwhile, on dry land, many clocks were set locally. When the sun was directly overhead, it was noon, but this changes slightly from place to place, so time differed from town to town. In the nineteenth century, railway networks came along, and a standard time, known as railway time, was brought in.

Get in the Zone

In 1884 the Royal Greenwich Observatory in London was chosen as the spot for the "prime meridian." This meant that the vertical lines of longitude on a map always started from 0° in Greenwich. It was also a handy spot, because at midday in Greenwich it was midnight on the other side of the planet in the middle of the Pacific Ocean. This meant that the fewest people possible would be affected by the change in date—along a line known as the international date line.

That's Atomic

The most accurate time instruments today rely on the regular vibrations of atoms. They are so accurate that one housed at the UK's National Physical Laboratory will lose or gain less than a second over the next 138 million years!

WHEN DAYS GO MISSING

Centuries ago, measuring the passage of a year was even more challenging than measuring the hours of the day. Many people relied on the sun or the moon to help them keep track of the days. In ancient Egypt, farmers based their calendar on the level of the river Nile. In ancient Rome the army calendar used the lunar month—the time it takes for the moon to pass through all its phases. The year was then divided up into 13 phases.

THIS YEAR'S JUST FLYING BY.

None of these systems worked very well. For example, the moon takes about 29½ days to go through its full cycle, so the Roman army's 13 cycles made 383½ days, because 13 × 29½ = 383½ days. This is longer than a true year, which is the time it takes the Earth to orbit the sun—a solar year.

Around the Sun in Just Over 365 Days

Why was it so hard to divide up a year? The answer is that the Earth orbits the sun in 365¼ days, minus 11 minutes, so it can't be divided evenly. Over time these extra ¼ days would build up, so that eventually your year would be so out of sync you'd be celebrating the new year in the summer.

Calendar Chaos

The ancient Romans had been noting the problem for many years, and in 45 B.C. Emperor Julius Caesar reformed the calendar. The problem was that it completely messed up the year. In fact, days slipped so much that Caesar had an extra 90 days added to 46 B.C. to correct the problem. No wonder it was called "The Year of Confusion"!

Lovely Leap Years

To stop the problem from happening again, the Romans decided to carry over the quarter days at the end of each year for three years. They then added an extra day at the end of every fourth year so that it became 366 days long. This became known as a leap year. However, the calendar still wasn't right—the extra quarter days were 11 minutes short! Over time these 11 minutes built up again, and by the mid-sixteenth century the calendar was 10 days off.

So in 1582 Pope Gregory announced that 10 days would be left out to make things right again. It was also ruled that a century year, such as 1600, 1700, or 1800, should not be a leap year unless it could be divided exact by 400. In some countries this correction didn't happen until much later. For example, in Britain the correction was done in 1752, when 11 days were left out. It is said this prompted riots by some people, who complained that the government had stolen 11 days of their lives!

NEVER MIND THE CALENDAR CHANGING—YOU FORGOT MY BIRTHDAY!

GETTING FROM A TO B

A calculation or a formula is very handy for mathematical information, but sometimes a picture is much more useful.

Take directions, for instance. It's simple enough to say, "Walk past Wishing Well Hill, through Bad Wolf Wood, and on the other side to Grandma's House," but what if you need to direct someone over a longer distance? It's easier to show this by drawing a map.

Sizing It Up

The only thing is, a map doesn't tell you about the distances between things or how big they are in relation to each other. What would be better is a map that is drawn "to scale."

This map has a scale, shown as a ratio, of 1:10,000. This means that 1 centimeter on the map is equal to 10,000 centimeters—or 100 meters—in the real world. You can use the scale on a map to help calculate distances.

To work out distances in a straight line, all you need is a piece of paper. Mark off the length of the scale—in this case, 1 centimeter—on your piece of paper, then place it on the map to calculate the distances between places. For example, it is 3.25 centimeters between The Wicked Queen's Castle and Grandma's House—on the scale of 1:10,000, this is 32,500 centimeters, or 325 meters.

If you want to measure a distance that isn't a straight line, you can use a piece of string to follow the curve of your route—around the base of Wishing Well Hill, for example. You can then measure the amount of string against the scale line on the map.

Scaling the Heights

Maps can even show how the land rises and falls, using contour lines. Mapmakers take points on a hill or valley that are the same height and join them up in a series of rings. The rings show the shape and relative heights of hills and mountains and the depths of valleys. Steep slopes are shown on a map with lines that are close together; gentler slopes have widely spaced lines. On this map, the hill rings showing Wishing Well Hill are very close together, so it is probably steep—it might be better to go around it!

Which Way Up?

To be able to read a map, you also need to know about direction. This is shown with an arrow to mark north pointing toward the top of the map. By using the map with a compass, you can make sure you are going in the right direction. It's useful because it works in all conditions.

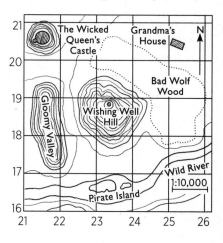

Groovy Grids

Most maps have horizontal and vertical "grid lines" that match the scale. In this map, each square on the grid is 1 centimeter wide. You know that the scale is 1:10,000, so you also know that each square represents 100 meters on the ground.

Grid lines help you to pinpoint an exact position by using a "grid reference." For example, you might ask your friends to meet you at 2317, but to find that exact spot, they'll need to know what the numbers mean. In this case, the first two numbers tell you how far *across* the map to look—the second two numbers tell you how far *up* the map to look.* Remind yourself with this useful saying:

Along the corridor and up the stairs.

Then everyone will know to meet on the riverbank north of Pirate Island—at the bottom left-hand corner of the square where lines 23 and 17 cross.

---

* The numbers across and the numbers up are sometimes known as eastings and northings.

X MARKS THE SPOT

You might need to pinpoint an even more exact location that lies between the grid lines—The Wicked Queen's Castle, for instance. In this case, you'll need to give a six-figure reference to help your friends find it. The number 216205 tells you the exact location of the castle.

The first two numbers tell you to go across the map to line **21**—the fourth and fifth numbers tell you to go up to line **20.**

You then need to imagine that each square is divided into 100 smaller squares, each 10 meters wide—10 across and 10 up. The third number shows the exact horizontal position: **6.** The sixth number shows the exact vertical position: **5.**

The Wicked Queen's Castle

This tells you that the castle is 60 meters east and 50 meters north of the corner of square 2120.

Clever Coordinates

If you know the coordinates, or grid references, of two locations, you can calculate the distance between them. In the map opposite, an explorer is shown at grid reference 212173—and he wants to get to Grandma's House, which is at 252204.

To work out the distance between the two, you can draw a right triangle over the map and use Pythagoras's theorem. This states that if the squares of the lengths of the two short sides are added, the result is the square of the length

90

of the longest side, or hypotenuse. (See pages 23–25 for more on Pythagoras.) So to find the distance to Grandma's House—the long side of the triangle—you first have to calculate the lengths of the two shorter sides.

Here the horizontal line shows you how far east Grandma's House is from the explorer's current location. The explorer is 20 meters east of grid line 21, and the house is 20 meters east of grid line 25. Each square is equal to 100 meters on the ground, so the horizontal line, **A,** is 400 meters long.

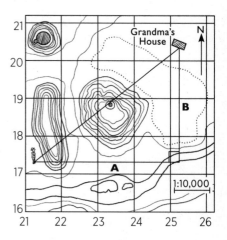

The vertical line shows you how far north Grandma's House is from the explorer. He is 30 meters north of line 17, and the house is 40 meters north of grid line 20, so the vertical line, **B,** is 310 meters long.

Now you can apply Pythagoras's theorem:

$$400 \times 400 = 160,000 \text{ meters, and}$$
$$310 \times 310 = 96,100 \text{ meters.}$$

If $a^2 + b^2 =$ the distance to Grandma's House squared, and these two figures added together make 256,100, then the house is approximately 500 meters from the explorer. (There is a special square-root button on some calculators marked "$\sqrt{}$" to help you work this out!).

DRAMATIC DATA

GREAT GRAPHS

The easiest way to organize a lot of information, or data, can be to use a diagram. Diagrams can communicate data clearly and easily and make it easier to understand—they can even be used to predict future measurements.

There are many types of diagrams that mathematicians use—graphs, bar charts, pie charts, and Venn diagrams are just a few. It's important to choose the right type of diagram for the information; otherwise, it won't make sense at all.

Plot It Out

A line graph can be used to plot information that changes over time. For example, this graph shows a boy's height from birth to the age of 18, plotted once a year:

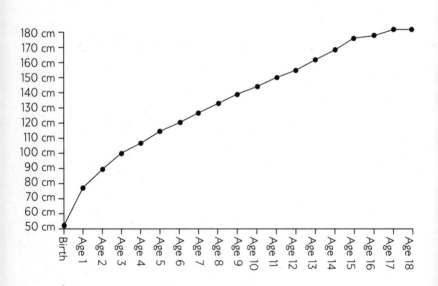

The bottom-left corner marks the starting point for the horizontal scale—also known as the x-axis—and the vertical scale—also known as the y-axis.

By joining the dots each year, the pattern of growth from birth to when the boy reaches adulthood can be seen. It's clear that the boy grew rapidly in his first year and that the rate of growth slowed as he neared 18.

Plotting the Future

In the line graph on the previous page, all the information you needed was included. What if you wanted to use a graph to find out what might happen next, though?

For instance, imagine a car is traveling along a road and a passenger is recording the time every 200 yards. She might end up with a table looking something like this:

Distance	Time (sec)
0	0
200	22
400	35
600	57
800	84
1,000	96
1,200	123

This information can be plotted on a graph with the time along the x-axis and the distance along the y-axis, like this:

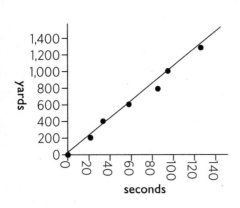

The speed of the car is changing all the time, so instead of joining all the points, a "best-fit" line has been drawn straight through as many points as possible. It gives you the car's average velocity (see pages 77–79).

You can use this best-fit line to calculate how long it might take the car to travel 1,400 yards. On the graph, you'll see that the line shows that if the car continued to travel at the same average velocity, it would take 132 seconds to travel 1,400 yards.

Tracking Trends

Graphs can also look at trends. A trend is the general direction in which something is moving. This graph shows the average life expectancy of U.S. residents from 1960 until 2010—over time the trend is that life expectancy is increasing.

If you draw a best-fit line through the trend line, it will show an average for the increase in life expectancy. In 1960 the average was 71 years—by 2010 it was just over 80 years. So in 50 years, life expectancy increased by 9 years.

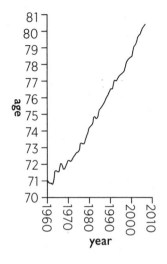

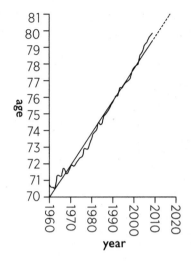

You can "extrapolate,"* or predict, how the information might go on by extending the straight line to show the next 10 years. Shown here with a dotted line, you can see that by 2020, the average life expectancy should be around 81.

Mistakes and Misunderstandings

Sometimes data on a graph can seem a little odd. Here is one for Cure-O, a hypothetical medicine that is meant to cure hiccups.

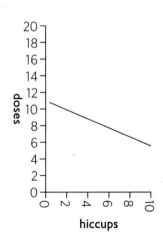

The first version shows the best-fit line for the data. It seems to suggest that if you took 20 doses of Cure-O, you wouldn't get hiccups at all.

*It's important to be careful with extrapolation—if you did the same thing backward, for example, with life expectancy decreasing by 9 years every 50 years, eventually the average life expectancy would be zero!

However, when you can see the individual doses, they are so widely scattered that there isn't really a trend at all.

In fact, if you take out one piece of data, the best-fit line changes, so it seems that the more Cure-O you take, the more hiccups you get!

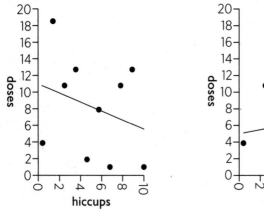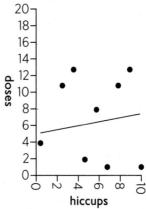

This goes to show how important it is to analyze data very carefully. It is easy to jump to conclusions once you start extrapolating, and that can lead to some strange results!

When you're plotting data on a graph, especially if you are using computer software to add best-fit lines, make sure that your results seem logical!

A MEAN SET OF NUMBERS

Graphs are just one way of grappling with data. They make up an area of mathematics called statistics, along with the methods below.

Averaging Out

If you catch a bus every day, one thing you might want to find out is how long the journey takes on average. The first thing you would have to do is note how many minutes several journeys take. Your data might look like this:

12, 11, 12, 12, 9, 9, 13, 15, 13, 13, 13, 13, 10, 10, 10

How could this data help you, though? To find out the average journey time, or "mean," as it is also known, you just add the numbers together, then divide the total by the number of numbers there are altogether. There are 15 numbers in total, adding up to 175, so:

$$175 \div 15 = 11.7$$

That tells you that your mean journey time is 11.7 minutes.

THIS PESKY ROADWORK IS WRECKING MY DATA.

What's in the Middle?

Rather than working out the mean average, you could work out the "median" instead. This is the data point in the middle of the group. The first thing to do is put the data in ascending order. There are 15 pieces of data, so the median is the 8th number—12—shown in bold:

9, 9, 10, 10, 10, 11, 12, **12,** 12, 13, 13, 13, 13, 13, 15

If there are an even amount of data points, the median is the middle two numbers added together, then divided by 2:

9, 10, 11, 12, 12, **12, 13,** 13, 13, 13, 13, 15

In this case, 12 and 13 are the middle numbers, which add up to 25, so the median is $25 \div 2 = 12.5$.

Mostly Modes

A third thing that can be found out is the "mode"—the data point that occurs most often.

9, 9, 10, 10, 10, 11, 12, 12, 12, **13, 13, 13, 13, 13,** 15

In this list the mode is 13, because it appears five times.

Distorted Data

What would happen if there was a huge delay one day and your bus journey took 80 minutes instead of 12? The total of the 15 numbers would be 243 instead of 175, so the mean would increase to 16.2 minutes. This skews your average journey time. This is where the mode and median are useful, because those numbers remain unchanged and can be used as a guide. This is why it's important to try to sort out your data before you come to any conclusions.

_____SORTING IT OUT_____

Imagine that the heights of 1,000 men have been measured and recorded.

A thousand numbers is a lot of data to handle. However, you could sort them into ascending order, then group them into ranges—140 to 144.9 centimeters, 145 to 149.9 centimeters, and so on. You can then count up the number of people in each range and plot them into a histogram, or bar chart, like this:

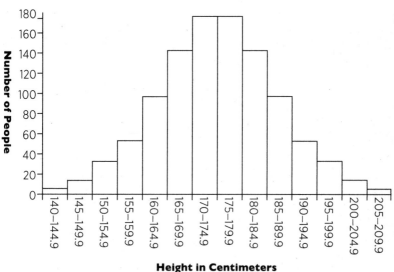

Height in Centimeters

This curved shape is known as normal distribution. In a normal spread of data like this, the mean, median, and mode are all in the same place, right in the middle.

By grouping the data like this, you can see that out of the 1,000 men, there were fewer who were at the shorter or taller ends, with more people grouped around the average.

⎯⎯ᴍⴸⴸⴸⴸⴸⴸ⎯ MR. VENN'S NUMBER TRAP ⎯⎯

Line graphs and bar charts aren't always the most effective way to show data. If you need to show the relationship between different sets of data, a Venn diagram is useful. It was named after John Venn, who came up with the idea in 1880. It consists of overlapping circles, showing different sets of data. The area where the circles overlap represents what the two sets of data have in common.

Give Peas a Chance

If you ask 20 friends if they like ice cream and if they like peas, you could put the results into a Venn diagram according to what they say. In this example, 18 friends have said they like ice cream—only 8 like peas as well, and the last 2 like peas but not ice cream:

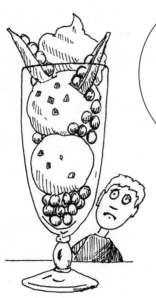

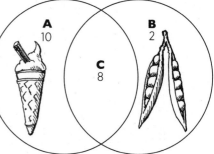

Circle **A** shows the friends who like ice cream. Circle **B** shows the friends who like peas.

Area **C,** where the circles overlap, shows the people who like both ice cream and peas— they are common to both sets.

A Survey of 100 People Says...

An inventor named Professor X has created a robot; he hopes that it will be extremely popular. Gizmoid can play soccer and tennis and look after your dog while you're at school. It seems like a brilliant idea—soccer and tennis are popular, and lots of people have a dog.

To make sure, Professor X does a survey, asking 100 people if they play soccer or tennis and if they have a dog. His data shows that 54 people like playing soccer, 32 play tennis, and 46 are dog owners. Only 3 people don't like soccer or tennis and don't own a dog.

This sounds very promising, but what's really useful about Venn diagrams is that they can show the relationship between more than two sets of data. If someone doesn't like any of the three things, they are shown outside the circles. So Professor X uses one to see how popular Gizmoid will be.

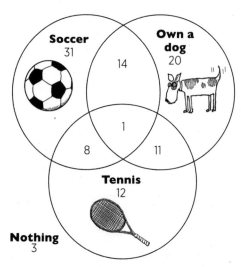

Immediately, Professor X can see the number of people who share each preference.

So, out of the 100 people Professor X surveyed, only 1 could make full use of Gizmoid!

A NICE SLICE OF PIE

Remember the creepy-crawly survey on page 74? The survey found 4 earthworms, 16 ants, 3 beetles, and 1 slug. Instead of just writing this data as a list, you could show it visually, using a pie chart, with different-sized slices of pie representing the information.

Pie Plotting

You might have a computer that can set out a pie chart for you, in which case, you can tap in your data, sit back, and relax. However, if you are drawing a pie chart by hand, your first task is to figure out how many degrees each type of creepy-crawly represents.

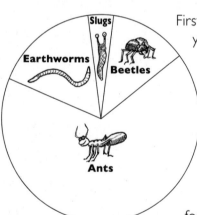

First, you'll need to add up all your creepy-crawlies to get the total—24. Then you need to work out the fraction of the total that each insect represents.

The 16 ants can be written as $^{16}/_{24}$, which is the same as $^2/_3$, or $0.6\dot{6}$ as a decimal. (See pages 12–13 for a refresher on this.)

Multiply $0.6\dot{6}$ by the 360° in a circle and you'll know how much of your pie chart needs to be given to the ants—240°. The earthworms need 60°, the beetles need 45°, and the slugs need 15°. Now you can divide your circle up into segments for each one. Note that if you add all these together, the total is 360°.

WHAT HAPPENS NEXT?

Sometimes it would be very useful to know what could happen in the future. For example, you might want to find out if it's going to rain over the weekend or if it will be sunny when you go on vacation. Even though it's tricky to know for sure, it is possible to say how likely it is that something will happen and express this using numbers.

Oddly Enough

A group of mathematicians called statisticians—people who are really into statistics—spend their lives working out the "odds," or chances, of something happening.

For example, the odds of winning the lottery are around 1 in 14 million—so you'd be very lucky to win it. The odds of being struck by lightning are 1 in 3 million—not so lucky, but still very unlikely!

What Are the Chances?

There is a trick to predicting the chance of something happening. You might want to know the likelihood of throwing a die and getting a 6, for instance. To figure this out, you would need to count the number of possible outcomes—how many different numbers can be thrown— and what fraction of those a 6 would be.

105

There are six possible outcomes when you throw a die, because it has six sides. However, only one of them is the one you want. This means that the chance of throwing a particular number is one in six.

If you want to find out the chance of throwing an even number instead of a particular number, you need to know how many there are out of the total. There are three even numbers, so the chance of throwing an even number is three in six—the equivalent of one in two. Of course, there is an equal chance of throwing an odd number, too.

How Predictable!

What statisticians are really interested in is probability. This shows, on a scale from zero to one, how likely it is that something will happen:

0	0.5	1
Impossible	**Even Chance**	**Certain**
It definitely won't happen.	There is as much chance that it will happen as it won't.	It definitely will happen.

For example, the probability of a pet goldfish dying at some point is 1, because a goldfish has to die eventually. You can also express this probability as a fraction or percentage: There is a 100% probability that a goldfish will die at some point.

To Be Precise

When you are talking about probabilities, it is important to remember that the way you express them can change the meaning. You need to be precise in your description of what is being measured. For example, the probability of a goldfish dying at a particular time, on a particular day, is very different from the probability of a goldfish dying at some point.

A Healthy Hypothesis

Remember Cure-O, the cure for hiccups on page 97. If it was a real medicine, the scientists who were testing it would have to be absolutely sure that it worked safely before it could be used to treat people.

Scientists testing a new medicine formulate what is called a hypothesis—a suggested theory—that they then try to prove or disprove. To do this, they gather and analyze data from tests to figure out whether the results are genuine or due to chance. They can then decide if the data is statistically significant or not. If the likelihood that the drug is effective is 0.99, or 99%, then they can confidently say that the drug is effective—unlike the imaginary Cure-O, which didn't work at all. Even if the drug works as a cure, tests will still need to be done to check that it is safe for humans to use!

MASTERLY MATH

TOP SECRET!

People have used secret codes to send messages for more than 4,000 years. Some codes are created by swapping letters of the alphabet or by using symbols, but many rely on mathematics. This science of codes and code breaking is known as cryptography.

Caesar's Cipher

More than 2,000 years ago the Roman emperor Julius Caesar devised a simple code called a transposition cipher. To make this type of code, you replace each letter in your message with another letter a certain number of places along the alphabet. Caesar usually went three letters along—so A became D, B became E, and so on.

Frequent Failure

The problem with cipher codes is that they are easy to break. This is because some letters come up more frequently than others and make patterns, which would appear in your cipher, too. For example, in English the most common letter is E. By looking at a cipher and figuring out which letter appears most, it is easy to see which letter has been substituted for E. If the letter H appears most, you can tell that it is most likely that each letter in the message has been changed for one that is three letters along the alphabet.

In 1465 an Italian named Leon Alberti published the first known "frequency table," which showed how often letters appeared and helped people to decode simple ciphers.

Is It Polybius?

A code that was a little more effective than the simple substitution cipher was created by a Greek man named Polybius, around 200 B.C. His code involved placing the letters of the alphabet in a grid.

	1	2	3	4	5
1	a	b	c	d	e
2	f	g	h	i/j	k
3	l	m	n	o	p
4	q	r	s	t	u
5	v	w	x	y	z

To encode a message with the Polybius square, you replace each letter with two numbers—the row number, followed by the column number. The letter H, for example, becomes 23. Try using the square to decode the following message (the solution is shown below):

54, 34, 45, 51, 15   13, 42, 11, 13, 25, 15, 14   24, 44, 43, 45, 35, 15, 42, 43, 31, 15, 45, 44, 23!

Unlocking the Square

It's still pretty easy to crack this code, once you've worked out that the most frequent letter, E, is represented by the number 15. However, you could add a cipher key—a word that changes the way the alphabet is arranged in your Polybius square—to make

	1	2	3	4	5
1	z	e	b	r	a
2	c	d	f	g	h
3	i/j	k	l	m	n
4	o	p	q	s	t
5	u	v	w	x	y

things trickier. This way, when someone sends you a message and you both have the same key—for example, "zebra," those letters would be put into the square first, followed by the rest of the alphabet. A code breaker would have to know the key, too, in order to fit the other letters in the right place and decode your message.

COMPLETING THE ENIGMA

An encryption machine that looked a little like a typewriter, known as the Enigma machine, was invented in the 1920s and was used by Germany during World War II. When a message was typed into it, a series of rotors scrambled it into cipher text to keep the communication secret.

The Enigma machine encryption was extremely complex, with so many possible combinations that many people believed it was unbreakable. What's more, the key code was changed daily at midnight. However, a group of people still worked tirelessly to try to crack the code. Alan Turing (see page 127) and his colleagues at Bletchley Park, in England, had an Enigma machine and an electromagnetic machine called a Bombe.

By putting in short parts of a message, known as cribs, that they had already been able to decipher, they could use the Bombe to disprove incorrect settings and find the correct ones. It is said that this top-secret project to crack the Enigma code helped to shorten World War II.

Computer Encryption

Information can now be encrypted and decrypted by computers in the blink of an eye—the Internet, banking details, e-mails—all encoded in more and more complex ways. Sometimes these codes are cracked by hackers, who have to use ever more ingenious ways of getting the information.

THE X FACTOR

An area of math called algebra can be used to calculate quantities when you have some of the information you need already. All you need to do is balance the equation—a calculation where both sides balance. Here's how.

Easy Equations

An equation, such as $x = 2 + 1$, might seem a bit odd at first glance, but there is an easy way of thinking of equations that will help you to work them out. Imagine a pair of old-fashioned scales with the equation sitting in them. Everything to the left of the equals sign must balance out everything to the right. The solution—the value of x—will have the same value as whatever is on the other side. You know that $2 + 1 = 3$, so if the scales are to balance, x must be 3 as well: $3 = 2 + 1$

Simple Subtracting

You might also come across equations such as $x + 1 = 2$. It looks a little trickier than the one above, but the rule that each side must balance still applies. To find out the value of x this time, you need to rearrange the equation, so that x is on one side of the equals sign, by itself:

$$x + 1 = 2 \text{ is the same as } x = 2 - 1$$

In other words, $x = 1$, so the equation is simply $1 + 1 = 2$.

Multiplication Muddles

Sometimes you might come across an equation where you have to multiply to get the answer:

$$x \times 3 = 10 \times 2 + 1$$

The way the right-hand side of the equation has been written is confusing. Is it $10 \times 2 = 20 + 1 = 21$, or $10 \times 3 = 30$? The way that mathematicians get around this is by using parentheses and working out what's inside first. So with

$$x \times 3 = 10 \times (2 + 1)$$

it is easy to see that $x \times 3 = 10 \times (3)$, which is 30.

Since $x \times 3 = 30$, to find x, you must divide the answer on the right by the 3 on the left, so if

$$x \times 3 = 30$$

it's the same as saying $x = 30 \div 3$. So x must be 10.

What Is Algebra For?

You may be looking at all these equations, wondering, What is the point of algebra? It might seem like a lot of bother to work out that $x = 1$ or $x = 10$, but this can come in useful in all sorts of subjects, as well as math.

For example, in physics, many important laws, such as Galileo's law of fall (see pages 78–79), are simplified as an equation. Equations are also used in chemistry, accounting, and even geography.

NICE ONE, SHERLOCK

. Logic is an area of mathematics that, like so many others, originated in ancient Greece. Logic is used to prove theories and ideas using strict, formal methods and systems.

Logically Speaking

Traditionally, logic was thought of as a branch of philosophy—the study of thought and ideas. One of the most famous examples of logic is a syllogism—a three-statement pattern. This one relates to the Greek thinker Socrates:

> All people are mortal.
> Socrates is a person.
> Therefore Socrates is mortal.

Logic doesn't tell you whether the statements are true, but it can be used to work out the idea of the statements and what must follow, if they are true. For instance, if it turned out that Socrates wasn't actually a person, but a statue, then the concluding statement would be false. It would be more correct to write the statements like this:

> *If it is true that*
> all people are mortal.
> *And it is also true that*
> Socrates is a person.
> *Then it must be true that*
> Socrates is mortal.

This is now definitely okay, whether Socrates is a person or a statue.

115

Applying Logic

A logical approach like this can be useful in mathematics. It makes you look at the structure of what is said and helps you to pinpoint problems with statements. You can also use it to make sure that anything you are working out is expressed correctly.

Logic Reinvented

In the nineteenth century, a mathematician named George Boole realized that the way in which logic was used in philosophy could also be used in algebra. This became known as Boolean algebra.

Boole's work was picked up in the twentieth century by three notable mathematicians—Friedrich Gottlob Frege, Bertrand Russell, and Alfred Whitehead. They wanted to show that by employing certain fixed rules, any mathematical theory could be proved with logic.

A MATH MACHINE

The first counting device was the abacus, which was in use as early as 4,000 years ago. Over the centuries, many other counting machines were developed—including Charles Babbage's difference engine, in 1822. Using a series of toothed wheels, it could make calculations and store data.

Switch On, Switch Off

The development of the kind of computers used these days was thanks to logic, and it started with the flick of a switch.

The flip-flop—not the type you wear on your feet—is a type of switch that can be used to store information. It was invented in the early twentieth century.

If a pulse of electricity was sent to the circuit with the flip-flop switch, it triggered a bulb to light up.

A second pulse, and the bulb went out again. It doesn't sound very impressive, does it? However, this allowed the trigger circuit to store information—the fact that a pulse of electricity had been received—in the simplest way possible.

It may not seem much of breakthrough now, and it didn't cause a great deal of excitement at the time either. However, by grouping lots of trigger circuits together, it was possible to store a sequence of pulses as a row of bulbs—some lit, others unlit.

Back to Binary

Here's where the flip-flop switch gets interesting. If you give the "On" and "Off" bulbs the numbers 1 and 0, it starts to look more and more like binary code (see pages 16–17). Here the lightbulb sequence is:

| ON | OFF | OFF | ON |

or, 1001 as a binary number. If you remember the columns of 1s, 2s, 4s, and 8s, that means one 8 and one 1, or 9:

8s	4s	2s	1s
1	0	0	1

DO YOU REALIZE WHAT THIS MEANS? WE'VE INVENTED DISCO LIGHTS!

By linking electronic circuits together, the sequences of lights in two rows of bulbs could be added together—in other words, they could be used to do calculations. Different circuits could also be set up to affect each other in different ways. For instance, they could be connected so that if either or both of two bulbs was lit, then a third bulb would be lit.

If neither were lit, the third would be off, too, like this:

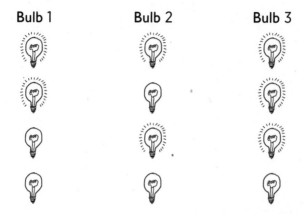

Bulb 1 **Bulb 2** **Bulb 3**

This type of circuit is known as a logic circuit. It represents the idea that if one or another thing is true, then a third thing is true too (see page 115).

Early Days

There is great debate about which computer was the first to be built. However, they were all enormous, filling entire rooms, were very expensive, and needed many people to operate them.

A Computer in Your Pocket

In the late 1960s, the development of an "integrated circuit," known as the microchip, enabled much smaller computers to be made. The home computer had arrived. The first ones had limited memories—as little as 16 KB— with programs loaded and stored on cassettes. Just four decades later, people can use their mobile phones, with many gigabytes of memory, to browse the Internet.

PROVE IT

Every new theorem must be proved before it can be accepted—and there are lots of ways to show proof.

Upside-Down Thinking

To find out if a statement is true, make the opposite statement. If that can be disproved, the first statement must be true. This is called proof by contradiction.

One of the best examples of proof by contradiction was a challenge solved by the Swiss mathematician Leonhard Euler. The challenge asked if it was possible to cross each of the seven bridges joining the four districts of Königsberg, in Germany, only once. To try every possible route would be extremely time-consuming, but Euler worked out the answer using proof by contradiction—all without leaving his comfy chair.

The first step was to assume that there definitely was a route where every bridge could be crossed just once. He quickly reasoned that for every bridge to be crossed once, there would have to be an even number of bridges. If there was only one bridge, you couldn't enter and leave a district without crossing it twice—if there were three, you could enter by one and leave by another, but the third would be left uncrossed. As there were seven bridges, Euler had proved by contradiction that there was no possible route where the bridges could all be crossed just once.

ARGUING IN CIRCLES

This unusual sign is an example of a paradox—something that is absurd or contradicts itself—it is impossible to do as it says. There are lots of paradoxes around. Some of them are just for fun; however, others have been used to test the limits of mathematics.

The Barber Paradox

"A barber only shaves the men who do not shave themselves."

British mathematician and philosopher Bertrand Russell used this statement to illustrate a paradox. It is absurd because it leads to a contradiction—if the barber were to shave his own beard, it would make him one of the men who shave themselves, so he couldn't be a barber.

The Incomplete Theorem

In 1931 an Austrian mathematician named Kurt Gödel proved his "incompleteness theorem." It stated that in any complex logical system, there must be some statements that cannot be proved or disproved, and that the system will have gaps that can't be filled. He also showed that mathematics is one of these systems. That means there must be some parts of math that can't be proved and others that will never be found. If they can't be found, no one even knows what they are—confusing!

MASTERS OF MATH

You've already heard about some of the work the finest mathematicians in history have done. Now here is a little more about just some of the people who have made mind-blowing discoveries over the centuries.

Pythagoras
(Around 580–500 B.C.)

A Greek mathematician who is often credited with coming up with the theorem named after him. Pythagoras and his followers believed that everything could be described with numbers. He also discovered irrational numbers. Little else is known about him and his followers, apart from the fact that they weren't allowed to eat beans or meat! (See pages 23–25 for more on Pythagoras.)

Euclid
(Around 325–265 B.C.)

Euclid was a Greek mathematician who wrote a 13-volume guide to geometry called *The Elements*. It is still regarded as a masterpiece of mathematical logic, and this branch of geometry is named Euclidean geometry after him, but little is known of Euclid's life.

Archimedes
(Around 287–212 B.C.)

The Greek mathematician Archimedes calculated pi with astonishing accuracy for the time. He also measured the area of geometric shapes and was an inventor.

Leonardo Pisano Fibonacci
(1170–1250)

An Italian mathematician—the Fibonacci sequence is named after him—who spent many years traveling. Back at home, he encouraged people to adopt the best methods he had found, such as using the digits 0–9, with his most famous work, the *Liber abaci* (Book of Calculation).

Galileo Galilei
(1564–1642)

Galileo was probably the first scientist to realize that the laws of the universe are mathematical. He helped prove that the planets go around the sun, explained the way that objects swing and fall, and used one of the first telescopes to study the night sky. His ideas were so radical that he was put on trial in 1634, and sentenced to house arrest for the rest of his life.

Johannes Kepler
(1571–1630)

Kepler managed to work out the mathematical laws that describe the way the planets move around the sun. He realized that the planets move in elliptical orbits and that their speed is not constant.

René Descartes
(1596–1650)

René Descartes not only invented modern graphs and coordinates but was also the first to use letters to represent unknown quantities in calculations.

GET LOST, FERMAT—I THOUGHT OF IT FIRST!

Pierre de Fermat
(1601–1665)

Fermat made discoveries in the theory of numbers, geometry, and optics, and so, unfortunately, did René Descartes. The two spent a great deal of time arguing about who was right and who was first. As Fermat never—for some odd reason—published anything, most of his work is known now through his letters. His "last theorem" is one of the most famous in mathematics and wasn't proved until 1995.

Sir Isaac Newton
(1642–1727)

COME OUTSIDE AND SAY THAT, DESCARTES!

Along with Albert Einstein, Sir Isaac Newton is regarded as one of the greatest scientists ever. He came up with mathematical laws to explain the way objects move and the way that gravity affects them. As a result, he was able to calculate the motions of the moon, planets, and comets. Many of his theories were published in his greatest work, *The Mathematical Principles of Natural Philosophy.*

Gottfried Leibniz
(1646–1716)

Leibniz developed several new areas of mathematics, including logic and "calculus"—a branch of mathematics related to algebra and geometry. He spent years arguing with Sir Isaac Newton about which of them had invented calculus, and even invented one of the first mechanical calculating machines.

Leonhard Euler
(1707–1783)

Swiss mathematician Leonhard Euler published many papers and books covering every area of mathematics from algebra to trigonometry. Even after he lost his sight in 1766, he continued to produce mathematical papers until he died.

Carl Friedrich Gauss
(1777–1855)

YES!

Gauss's talent for mathematics emerged at a very early age. He made major discoveries in many areas, including statistics, the theory of numbers, and geometry—he also correctly calculated the next sighting of the dwarf planet Ceres.

Albert Einstein
(1879–1955)

The German American physicist Albert Einstein often complained that he was no good at math. In spite of this, he still came up with a new mathematical theory of gravity, which improved on Sir Isaac Newton's. He also showed how time can slow down or speed up, and came up with the most famous equation in the world, $E = mc^2$. Not bad for someone who said he was no good at math!

Kurt Gödel
(1906–1978)

As well as showing that mathematics contains unfillable gaps and unprovable statements, Kurt Gödel also explored the idea of time travel.

Alan Turing
(1912–1954)

During World War II, Germany used the Enigma machine to encrypt messages (see page 112). Drawing on the work of Polish mathematicians, such as Marian Rejewski, Turing and his colleagues finally cracked the code, even though it had been thought to be unbreakable. This work helped to bring the war to an end.

Srinivasa Ramanujan
(1887–1920)

Indian mathematician Srinivasa Ramanujan showed an incredible talent for mathematics, even as a child. By the time he was in his teens he was coming up with his own theories. Sadly, he died when he was only 32, but in that time he had written many papers.

Sir Andrew Wiles
(1953–present)

A mathematics professor at Princeton University, Wiles finally proved Fermat's last theorem in 1995. It took 10 years to work out and 200 pages to write down.

MATHEMATICAL GENIUS

A NUMBER OF NUMBER TRICKS

If you want to impress friends with your magical abilities and impress your teachers with your startling mathematical skills, these next few pages are for you!

There is a selection of amazing number tricks, mathematical shortcuts and tips, and a collection of mathematical mnemonics to help with memorizing handy bits of math.

Trick One: Marvelous Mind Reading

For this trick, all you need is a friend and a calculator.

1. Ask your friend to think of a three-digit number where the first and last digits are at least two numbers apart. For example, 923 would be fine, because 9 and 3 differ by six—253 wouldn't work because 2 and 3 are too close.

2. Now ask your friend to reverse his number, so for example, 923 becomes 329.

3. Next ask him to subtract whichever is the smaller number from the larger number, using the calculator if he wishes. 923 − 329 = 594.

4. Then ask him to reverse his answer, so 594 becomes 495. Note that, at this point, the pair of numbers should always be 594 and 495.

5. Finally, ask him to add both answers together—so 594 + 495. Hold up your hand and say, "Don't tell me. I know the answer!"

6. Pause for effect, then say, "Is your answer 1,089?" and wait for your friend to gasp in amazement.

7. Lastly, smile smugly and leave them to wonder, "How did you do it?"

Trick Two: Secret Assistant

This trick needs a secret assistant in the audience who is fairly good at mental arithmetic and who can also act surprised on cue. You will also need at least three more people to impress. You'll need to make a few preparations to make sure this trick works well. Here's what to do:

The preparations

Before your audience is in place, agree on a number with your assistant that is larger than 6,000 to go into the envelope in step **1.**

Your assistant will need to be able to subtract the numbers decided on in steps **3, 4,** and **5** from the agreed number—this will be where mental arithmetic comes in.

In step **6,** she will have this answer ready to use as the year she has "chosen," and can jot it down with the number you both agreed on written underneath.

The performance

1. Begin your trick by saying, "I'm thinking of a number," in a dramatic voice. With a flourish, write down the number that you agreed with your assistant—7,684, for example, and seal it in an envelope.

2. Leave the envelope on view for the audience for the rest of the trick—make sure that they can see it clearly.

3. Ask a member of the audience which year she was born in and write her answer on a large sheet of paper—1973, for example.

4. Now ask another person for a memorable year in her life, 2009 perhaps, and write that down, as well.

5. Ask a third person to give you a notable year from history—1776, for example—and write that down, too.

6. Tell your audience that you will make the trick harder by asking someone to choose another year—past or future—and that she should write it down at the end of your list, but so that you can't see it. Make sure you choose your secret assistant for this job! She will need to write down the number she has worked out during the trick. Using the numbers from steps **3, 4,** and **5** as examples, her year should be 1926 because 1,973 + 2,009 + 1,776 + 1,926 = 7,684.

7. Now ask your assistant to "add up" all four numbers in your list ready to announce the total. While she is adding up, pass the sealed envelope to another member of the audience.

8. When your assistant announces that the number is 7,684, your audience will be amazed when they open the envelope and discover that the numbers are the same!

_____TIPS AND SHORTCUTS_____

When you are working with numbers, there are often useful shortcuts that can help you to find a speedy solution. Here are just a few to get you started.

Dividing by Five

It can be tricky to divide by 5, unless you learn all the answers by heart, but dividing by 10 is much easier. Any number that you need to divide by 5 can be divided by 10 and then doubled to find the answer, because $5 = 10 \div 2$.

For example, 90 ÷ 5 might seem difficult at first glance, but 90 ÷ 10 is much more straightforward—the answer is 9.

Then you can easily double the answer to get 18, so:

$$90 \div 5 = 18$$

Try it for a larger calculation: 280 ÷ 5, for example—and it still works. 280 ÷ 10 = 28, 28 x 2 = 56, so 280 ÷ 5= 56.

Is It Divisible?

Want to guess how a number can be divided? There are some simple tricks to help you work things out.

If a number is divisible by…

… 3, its digits will add up to a multiple of 3. For example, you can tell that 8,904 is divisible by 3 because 8 + 9 + 4 comes to 21.

… 4, the last two digits will either be two zeros, or will be divisible by 4.

… 10, it will always end in zero.

Easy Elevens

To multiply a number by 11, first multiply your number, for example 425, by 10, so 425 x 10 = 4,250. Then add the first number to your answer, so 4,250 + 425 = 4,675.

Nine More, Please

When you're multiplying by 9, there's an easy trick to get to the answer quickly.

1. Spread out your hands with your palms facing you.

2. Bend the finger or thumb corresponding to the number you want to multiply 9 by. For 9 × 4, bend the fourth finger on your left hand.

3. Count the thumb and fingers to the left of the bent finger ... 3, in this case. This is the number of 10s in the answer.

4. Now count the fingers and thumb to the right ... 6, in this case. This is the number of ones, so the answer is 36.

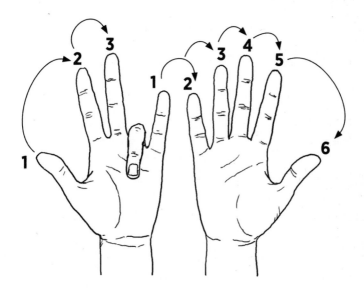

Quick Calculations

Adding up a list of numbers can be daunting, especially if they are not whole numbers. Take this list:

$73.4 + 98.0 + 61.9 + 83.1 + 3.6 + 56.6 + 64.1 + 50.0 = 490.7$

If you only want a rough answer, then rounding them all to the closest digit before adding makes less difference than you might think:

$73 + 98 + 62 + 83 + 4 + 57 + 64 + 50 = 491$

If you really want to save time, you could round them to the nearest 10 for a rough total, which is still less than 3 percent off:

$70 + 100 + 60 + 80 + 0 + 60 + 60 + 50 = 480$

How Old Am I?

This cunning trick will let you guess someone's age, using nothing more than the powers of calculation. This trick will work on anyone over the age of 10, so it's perfect for amazing your teachers or your parents. Here's how to do it:

1. Ask your chosen person—maybe his name is Joe— to multiply the first number of his age by 5, but to keep this number secret from you. For example, if Joe is 23, he would multiply 2 x 5, to get 10.

2. Ask Joe to add 3 to the number he has just calculated. He would add 3 to 10, to get 13. All the calculations should be done secretly, so you don't hear what numbers he is using.

3. Ask Joe to double the number he has just worked out. So, for example, he would multiply 13 x 2 to get 26.

4. Next up, Joe should add the second digit of his age to the figure he got in step **3.** He would add 3 to 26, giving him 29. Ask him to tell you this number.

5. Now all you have to do is subtract 6 from the number that Joe has told you, and voilà, this new number will be his age.

This trick should work no matter how old the person is— just like magic!

137

MATHEMATICAL MNEMONICS

A mnemonic is something that can help you remember things. Sometimes it is a rhyme, or it can be an acronym—a word made up by abbreviating the first letter of each word in a list. PIN, for example, is an acronym for Personal Identification Number.

MY NEW ELEPHANT MUNCHES ONLY NEW IMPORTANT CARS.

How Many Days?

Whether you're counting down the days until your birthday, or trying to calculate the day you were born, this rhyme comes in handy:

> 30 days hath September,
> April, June and November
> And all the rest have 31
> Except for February alone
> Which has 28 days clear
> And 29 in each leap year.

How I Wish I Could Remember Pi!

Remembering pi to more than a couple of decimal places might be less difficult than it seems at first. All you need is a rhyme or sentence where each word has the corresponding number of letters to the numbers in pi. For example, "How I wish I could calculate pi before lunch, for lunch includes fantastic rounded dumplings," gives you pi to 14 places:

3.14159265358979

A Fraction Reminder

In a fraction (see pages 27–28), which number is the numerator, and which is the denominator? Use this as a reminder:

NUmerator Up,
Denominator Down

Mean, Mode, or Median?

When you're dealing with data and averages (see pages 99–100), how do you remember the difference between mean, mode, and median? It's quite muddling.

Mode and most both have four letters and both start with "mo." You can remind yourself with a sentence like this:

"There once was a mean average. He had a median caught in the middle. The mode appeared most."

Compass Directions

To make good use of the maps on pages 87–91, you'll need to know which way north, south, east, and west are. Use one of these mnemonics to remember them clockwise from the top, where north is:

"Never Eat Shredded Wheat" or
"Never Eat Slimy Worms."

INDEX

Reader's Digest Books for Young Readers

I Wish I Knew That

This fun and engaging book will give young readers a jump start on everything from art, music, literature, and ancient myths to history, geography, science, and math.

STEVE MARTIN, DR. MIKE GOLDSMITH, AND MARIANNE TAYLOR
978-1-60652-340-7

i before e (except after c)
The Young Readers Edition

Full of hundreds of fascinating tidbits presented in a fun and accessible way, this lighthearted book offers kids many helpful mnemonics that make learning easy and fun.

SUSAN RANDOL • 978-1-60652-348-3

Write (Or Is That "Right"?) Every Time

Divided into bite-size chunks that include Goodness Gracious Grammar, Spelling Made Simple, and Punctuation Perfection, this book provides quick-and-easy tips and tricks to overcome every grammar challenge.

LOTTIE STRIDE • 978-1-60652-341-4

I Wish I Knew That: U.S. Presidents

Starting with who is qualified to be president, what a president does, and how the president works with the rest of the government, this book, written just for kids, quickly turns to the fascinating profiles of each of the 43 presidents. There are also sidebars filled with fun and unusual information about our leaders.

PATRICIA A. HALBERT • 978-1-60652-360-5

Other Titles Available in This Series

I Wish I Knew That: Geography
Liar! Liar! Pants on Fire!
I Wish I Knew That: Science

Each book is $9.99 hardcover

For more information and to download teacher's guides, visit us at RDTradePublishing.com

E-book editions also available

Reader's Digest books can be purchased through retail and online bookstores. In the United States books are distributed by Penguin Group (USA), Inc. For more information or to order books, call 1-800-788-6262.

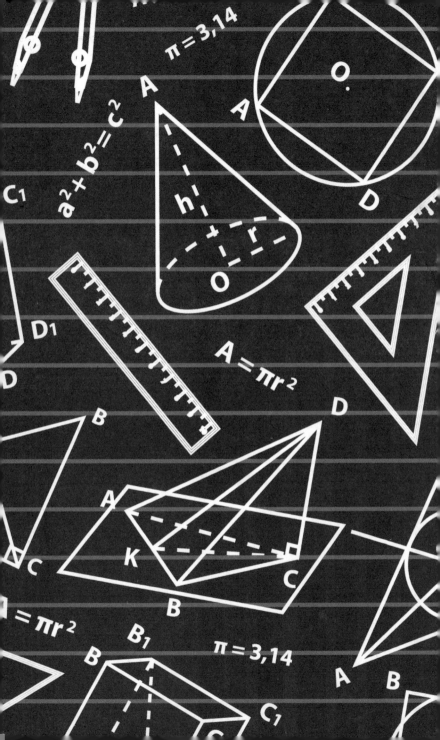